D0915622

FAR-FETCHED FACTS

Inside Technology
edited by Wiebe E. Bijker, W. Bernard Carlson, and Trevor Pinch

A list of books in the series appears at the back of the book.

FAR-FETCHED FACTS

A Parable of Development Aid

Richard Rottenburg

translated by Allison Brown and Tom Lampert

The MIT Press
Cambridge, Massachusetts
London, England

Originally published in German under the title *Weit hergeholte Fakten. Eine Parabel der Entwicklungshilfe* (Lucius & Lucius, 2002)

MIT Press books may be purchased at special quantity discounts for business or sales promotional use. For information, please email special_sales@mitpress.mit.edu or write to Special Sales Department, The MIT Press, 55 Hayward Street, Cambridge, MA 02142.

This book was set in Scala and Scala Sans by Graphic Composition, Inc., and was printed and bound in the United States of America.

Library of Congress Cataloging-in-Publication Data

Rottenburg, Richard.
[Weit hergeholte fakten. English.]
Far-fetched facts : a parable of development aid / Richard Rottenburg ; translated by Allison Brown and Tom Lampert.
  p.   cm.—(Inside technology)
Translated from the German.
Includes bibliographical references and index.
ISBN 978-0-262-18264-5 (hardcover : alk. paper)
  1. Economic assistance—Developing countries. 2. Social change—Developing countries. I. Title.
HC59.7.R648   2009
338.9109172'4—dc22

                                                2008041180

10 9 8 7 6 5 4 3 2 1

Contents

Acronyms

AOD   Agency for Overseas Development (fictive)

FC   financial cooperation

GNI   gross national income

IMF   International Monetary Fund

l/c/d   liters per capita per day

MDC   Ministry for Development Cooperation

MIS   management information system

NDB   Normesian Development Bank (fictive)

NGO   nongovernmental organization

NUWA   National Urban Water Authority (fictive)

O script official script

OECD   Organization for Economic Cooperation and Development

OIP   Organizational Improvement Program

pm   person-month (the work capacity of one person per one month)

S&P   Shilling & Partner (fictive)

TC   technical cooperation

U script unofficial script

USAID   United States Agency for International Development

UWE   urban water engineer (fictive)

WHO   World Health Organization

WMS   water management system

Prologue

The end of colonialism marked the advent of a new kind of expert and a new kind of global network of organizations. The ongoing concern of this new field of expertise and this new epistemic community has been to initiate economic and social development in the poorer countries of the southern hemisphere. The key problem in this endeavor lies in establishing objectivity between different frames of reference. Scholars studying the field of development cooperation can either collaborate with the development experts in order to construct more objective representations of one particular development problem or another, or they can mingle with these experts in order to examine their representational practices and the consequences of these practices. In this book I have attempted to do the latter and only the latter.

The Object of Study

In the late nineteenth and early twentieth centuries, a process that had begun approximately four hundred years earlier came to an end. A few nations located primarily along the northern Atlantic coast divided up almost the entire non-European world into empires. These empires then collapsed in the two decades following the Second World War. Although the particular dynamic that triggered the colonization of the non-European world did change with the end of colonialism, it could no longer simply be undone. The postcolonial version of this dynamic is called *development*. As prescribed by capitalist and (until 1989) socialist industrialism, that is, by economic and technical-scientific progress, development was supposed to result in improved standards of living.

This model defined progress through *modernization* as the sole proper path and in doing so denounced local ways of living as backward. According to this distinction, approximately two-thirds of the world's population remains underdeveloped. The most outrageous thing about this distinction is its universal acceptance, often even by those classified as underdeveloped. People who view themselves in these terms are ashamed of their language, their dress, and their customs and convictions—or at least those of their ancestors. And if they want to make something of themselves, they believe that they can do so only through a belated socialization at one of the elaborate educational institutions set up in the developing countries. The knowledge they acquire at these institutions will have little or nothing to do with their specific social and cultural environment, for the simple reason that this world is regarded as an impediment that is best left behind. Instead, the development model is taught, and this has significant consequences for the perspective of elites in these countries.[1]

What we call development is by no means a uniform process. Although it is possible and even necessary to determine the dynamics of development according to phases, continents, and countries, a uniform pattern of actions and interpretations can nevertheless be identified in ideal-typical terms. This pattern includes the following elements: a *society* that is labeled "underdeveloped" and thus by definition has been unable to achieve the goal of development on its own as a kind of involuntary social transformation; a class of *elites* who believe they have been called upon to modernize their own society; a *model* that promises to overcome underdevelopment; international *experts* who assist local elites in implementing this model; and a global network of formal organizations that engage in and finance the process of development. Here we can distinguish between donor and recipient organizations as well as between national and multinational organizations; all of these organizations, however, populate what is called an *organizational field* characterized by distributed agency. Last but not least, the discourse of this development arena is supported by the global *hegemony of the Western worldview*.[2]

If we locate the beginnings of development policy in the 1960s, efforts to promote development have now entered their sixth decade. The term "development aid" has not been used in official discourse for many years, having been replaced by the term "development cooperation," as it is called in international jargon. This new term implies that one party (which has more and is farther along) transfers something to another party (which has less and has not come so far). The objective of this transfer is that the poorer party will at some point in time no longer be dependent on this assistance because it has made up the difference. That is the simple quintessence of this process.

Over the course of decades, however, the hopes and the certainties of development discourse have steadily been deflated. This is especially true in sub-Saharan Africa, which constitutes the focus of this book. Parallel to the continuous failures of this kind of belated modernization and in part independent of them, there has also been a steadily increasing uncertainty within the so-called donor countries, which has culminated in the slogan "learning from other cultures." Although this kind of self-doubt has been more evident among interpretive elites in the West than among their decision-making counterparts, it has nevertheless become so prevalent and powerful that the US–European development model is now articulated in public discourse only in code and with a sense of shame. In 1951, it was still possible to assume the following tone:

There is a sense in which rapid economic progress is impossible without painful adjustments. Ancient philosophies have to be scrapped; old social institutions have to disintegrate, bonds of caste, creed, and race have to burst; and large numbers of persons who cannot keep up with progress have to have their expectations of a comfortable life frustrated. Very few communities are willing to pay the full price of economic progress. (United Nations, Department of Social and Economic Affairs, Measures for the Economic Development of Underdeveloped Countries, 1951)[3]

In 1996 the language employed is rather different: "[W]e in the West cannot tell people how they should develop. They are intelligent enough to be able to decide this for themselves and for their children" (Wolfensohn, President of the World Bank).[4]

The collapse of socialism, which is usually dated with the fall of the Berlin Wall in 1989, left the Western market model as the sole remaining societal form. This lack of alternatives and competition meant that earlier considerations about the prospects of the development model could now be thought through more rigorously than had seemed possible during the era of the cold war. It was recalled that the model in fact undermines its own natural basis. From this vantage point, a belated modernization outside of the US–European world is something to be prevented by all means possible. The older insight that the Western market model systematically renders human beings superfluous also became increasingly evident. Today the integration of the entire Third World population into an unaltered capitalist system of production can no longer be seriously regarded as a desirable alternative.

The development model, however, has lost its luster primarily because—with some exceptions in Asia and South America—poverty has increased rather than decreased since the beginning of development cooperation. The end of colonialism and fifty years of development aid have not resulted in greater security, dignity, or prosperity for the majority of the population in sub-Saharan Africa. On the contrary, the description of the Congo in a local newspaper article from 2000 could also be applied to many other locations: "Independence has turned out to be a nightmare."[5] Although the sites of postcolonial nightmares should not be turned into symbols for all of sub-Saharan Africa, it is also essential that we not shy away from placing these nightmares in the proper context for fear of making false generalizations. The overarching questions that arise here are: Why has this happened and why does it always seem to repeat itself? It will certainly be necessary to continue looking for answers to these questions on different levels and from diverse perspectives.

In the present book, however, I have limited my analysis the role of technologies of inscription and representation in development cooperation as an organizational process. In doing so, my focus will be neither the actual development at a particular location nor the construction of a theoretical conceptualization of development. I concentrate instead on the practices of organizing development cooperation that occur in *interstitial spaces*—neither entirely where the model ostensibly originated nor entirely where it is supposed to be implemented.

The Organizational Field of Development Cooperation

The global organizational field of development cooperation is populated by a plethora of organizations possessing a diversity of legal forms. Large international organizations were established to deal with the cataclysms and consequences of the Second World War; later, in the 1960s, these organizations gradually began to assume primary functions in the domain of international cooperation and development work. The two most important and highly complex organizational clusters here are (1) the World Bank Group (with five suborganizations) and the International Monetary Fund, which together are also known as the Bretton Woods Organizations, and (2) the United Nations with its diverse special organizations, of which the United Nations Development Program (UNDP) is centrally concerned with issues of development. All industrial nations of the northern hemisphere also have national organizations dedicated to the work of development cooperation.

In 1960, under the leadership of the United States, several member states of the Organization for Economic Cooperation and Development (OECD) founded within the OECD the so-called Development Assistance Group, which was renamed shortly thereafter the Development Assistance Committee and which has served as an arena for negotiations on international agreements regarding bilateral development cooperation.

Outside the realm of official development assistance, a variety of nongovernmental organizations (NGOs) and foundations have sprouted from the ground of the organizational field of development, in particular since the last globalization wave in the early 1980s. These organizations and foundations frequently claim to deal with the real problems of people living in the poor countries of the South more effectively than state development institutions can. In many cases there is a complementary relationship between the large programs of official development assistance and small NGO projects, which are in part commissioned and financed by these larger programs.

At the United Nations Conference on Environment and Development in Rio de Janeiro in 1992, most nations of the world voted to adopt the Agenda 21 program. According to this program, wealthy industrialized nations should strive to allocate 0.7 percent of their gross national income (GNI) to development cooperation (only the Scandinavian countries have to date actually done so). The United States never agreed to adopt this goal and in fact allocates only 0.17 percent of its GNI to development, placing it twenty-first among the industrialized nations. Measured in absolute terms, however, the United States remains the largest provider of official development assistance, followed by the United Kingdom, Japan, France, and Germany.[6] Donor nations have made an increasing portion of these funds available to international organizations, while the remainder is used to implement bilateral cooperation. Development cooperation involves enormous sums of money—for example, in 2006 the twenty-three OECD nations that belong to the Development Assistance Committee allocated 104.4 billion dollars in official development assistance, which corresponded to 0.31 percent of the combined gross national income of these countries; even more overwhelming, the West spent $2.3 trillion on foreign aid over the first five decades of development cooperation. However, the issue looks rather different if these figures are compared with the significantly greater sums that OECD countries spend to protect their own markets and workplaces from products produced in precisely those nations receiving development monies, as well as the sums that these nations could earn if trade barriers were removed.[7]

Since the 1990s there has been a shift from systematic efforts at establishing structures for belated and sustainable development to selective "humanitarian interventions" triggered by crises and catastrophes. This shift appears to be tied to a rather opaque combination of the following tendencies: The post-cold-war era has been marked by a reconfiguration of the world order and a universal implementation of neoliberal markets and neoliberal regimes of governance under the leadership of the sole remaining superpower, the United States. At the same time, the discourse of human rights has gained in political significance and been linked to a health discourse that seeks to place the demand for health care on a par with universal human rights. This appears to have given rise to a historically new form of dominance, one that establishes its legitimacy through the presupposition of catastrophes and states of emergency, which then allow it to engage in humanitarian interventions addressed at endangered bodies. A new constellation of military, private enterprise, civil society organizations, and health care has emerged as a result, which calls into question the structural principles of modernity.[8]

The present study focuses on the decade of the 1990s, that is, the last years of the "developmentalist era," when the paradigm of sustainable development as a project that sovereign nation-states undertook with international assistance came to an end. While the organizational field of development cooperation as such is thematized in this book, I have chosen to focus on a project that was financed by the development bank of a European country and implemented by a private consulting firm under the supervision of African project-executing agencies. This approach has allowed for a detailed investigation of only a particular section of the organizational field; however, my primary interest and focus are elementary questions that necessarily play a key role in all national variants of development cooperation.

In several donor countries, the responsibility for development cooperation is relatively closely tied to the respective governmental authorities and their political parameters. In the UK, for example, development cooperation is steered to a large extent by the Department for International Development (DFID, formerly Overseas Development Agency, ODA). In the United States, the National Security Council (NSC) coordinates all activities in the realm of official development cooperation, although the actual work is done by United States Agency for International Development (USAID), which is in principle independent yet closely affiliated with the State Department. In other countries, especially Germany, governmental authorities for development cooperation exist, but they define only general guidelines, leaving the implementation largely to independent organizations that engage in this work on a commission basis.

All the different organizational variants of the development enterprise are concerned with a simple but extremely tricky problem. If I lend money to someone who then uses it legally to earn more money and I in turn am paid interest, then this interest is sufficient proof that it was a good investment. If, however, I lend money belonging to a third party to someone who proves unable to make a profit with it, then I need a different kind of proof to demonstrate that this was nonetheless a good investment. In this case it is necessary to show in detail how the money was actually used. In the global organizational field of development cooperation, participants thus have to document the activities of projects that have been initiated by development investments. This book investigates the techniques and technologies of these representational practices of documentation.

Locations, Times, and Actors

The material for this study—like the material for any ethnographic study—could be collected only at specific points in time, at concrete locations, and on the basis of real events. Between 1978 and 1983, I spent more than three years doing field research in the Nuba Mountains (Sudan). During this time I became acquainted with various development projects from the perspective of the local farmers—projects which, for example, sought to convince these farmers to adopt the oxen plow. It was only later, between 1990 and 1998, that I investigated development projects from an internal standpoint. Initially I worked during semester breaks at my university as an organizational analyst for a development bank. My work in this capacity was centered on five development projects in four different African countries (Gambia, Ghana, Tanzania, and Lesotho). The pattern was always the same: The projects had been impeded by problems that from the perspective of those responsible could be classified as neither technical nor economic. My task consisted in identifying the "sociocultural reasons" why project objectives had not been achieved. All of these projects were already in advanced stages or had been completed, and all of them involved the organizational embedding of large technical systems that were intended to provide a public infrastructure (transport and water supply). This role acquainted me primarily with the work of project-executing agencies and consultants from the perspective of the financier.

In 1992 I worked for six months as a trainee at the same development bank and became well acquainted with this institution from the inside as well. The bank expressly requested that I observe its work and report on my observations, which I did. In the meantime I also worked for a consulting firm, evaluating its work in the port of Maputo (Mozambique). Finally, for more than two years (1996 to 1998), during free periods at my university I worked as a freelance staff member for a consulting firm that implemented an Organizational Improvement Project (OIP) at the waterworks of three Tanzanian cities (Arusha, Moshi, and Tanga) on commission of this same development bank. The development bank had requested that this project employ an anthropologist to better incorporate the "sociocultural aspects" of the undertaking. In this role I learned the perspective of both consultant and project-executing agency in detail. In order to supplement this ethnographic material, I conducted interviews in January 1998 at the ministry responsible for development cooperation. This book, in other words, is based on nineteen months of multisited field research at nine organizations located in five African countries and at one European development bank. I participated in all of the conversations, meetings, and negotiations investigated in detail in the present text and kept a meticulous diary on them in written form and on audio tape.

Although the book draws upon my entire experience with development cooperation and the cumulative results of that work, my explicit argumentation here refers solely to material from the final development project in Tanzania and thus corresponds to a classical ethnographic case study. Nevertheless, the case depicted in this book has been fictionalized. I decided to do this for several reasons. In contrast to an anthropological study of laboratory practices focusing on the analysis of protein communication, for example, the issues addressed here inevitably raise pressing evaluative questions—questions that can ostensibly be navigated by means of common sense, that appear to be primarily political and moral in nature, and about which everybody feels called upon and entitled to offer critical comments. Readers are almost instinctively drawn to ask the following questions: "Why does this nonsense continue?"; "Who is in charge here?"; and "How can it be done better?" Identifying real actors would only encourage readers to latch onto questions of individual responsibility. The fictionalization of my account was intended to counter this. I wanted to direct attention away from the strengths and weaknesses of specific real actors and toward the significance of general structural principles and the contingencies of the mundane practices of the development world.

After composing an initial text draft that included the names of real persons, institutions, and locations, I became convinced that the fictionalization of the report was also a question of decency. It seemed to me intrusive and offensive to publish a text in which real human beings were so ruthlessly exposed, even if they had previously given their approval for the study. One alternative would have been to reduce the vitality of the final text so that the people involved appeared in a paler, more muted light. This, however, was incompatible with my intention of composing an ethnographically thick and vibrant text, in which contingencies and personal idiosyncrasies played a significant role. For this reason I decided to fictionalize my account. In my depiction I have consistently sought to impute only good intentions to all of the figures involved and to place them in the best possible light—so as not to induce readers to wander onto the side stage of individual responsibility, which has little or no direct relevance to the issues investigated here. Nevertheless I cannot be certain that no one will feel insulted by what I have written. While this does remain an inevitable possibility even with fictionalizations, I contend that it makes an enormous difference over the course of the years both to the people concerned and to those who know them whether real names have been included or not.

My decision to fictionalize was also based on a psychological consideration. I knew that a substantial number of readers of this book would be actors in the global field of development cooperation. When these readers got hold of a report that prominently identified a particular development bank by name, they would be confronted with one of two problems. They might be affiliated with this bank and thus feel violated in terms of loyalty or their own self-esteem. They would also be unable to simply leave the book on their desks and discuss it with their colleagues during their lunch break. Or they might be affiliated with a different development bank and be misled through the identification of this particular bank into attributing the problems analyzed in the book to the identified competitor, which could prevent them from gaining any additional insight that my analysis might offer. In the years since the book was published in Germany, I have received dozens of emails from actors in the field who have thanked me for the accurate analysis and have affirmed my decision to fictionalize the account.

Finally—and perhaps decisively—this book is also intended as a contribution to what could be called experimental ethnographic writing. In the era after the end of critique, grand narratives, and utopias, scholars must nevertheless continue "to speak truth to power" and to seek a new language with which to do this. Those who conduct ethnographic studies of powerful organizations must, on the one hand, engage the practices and conceptions of those organizations with the respectful affection of an anthropologist. On the other hand, they must not conceal or gloss over things that might have negative consequences. In my experimental writing I have attempted to achieve this balance through fictionalization.

All of the characters in the present text have been given fictional names and are literally figures in a play. They do not depict any real, existing people but are constructed from the cumulative characteristics originally belonging to the various people I met during my tenure in the field of development cooperation. They wear the masks and play the roles prescribed by the script, and yet at the same time they perform with the maneuvering room that I found typical of the development arena. At issue are not their individual capabilities, honesty, or good intentions; rather, it is presumed that all figures possess the normal competency required for the roles they play. If their interactions do not bring about the desired results, this cannot be traced back to the failing of one or another of the actors.

The events of my narrative are set between July and December 1997. The retrospective accounts provided therein expand this time frame to encompass all of the 1990s. The main institutional actors and settings of the narrative are: (1) the Normesian Development Bank (NDB), the development bank of a fictional European country called Normland with its seat in the fictive city of Urbania; (2) Shilling & Partner (S&P), a small consulting firm located in Mercatoria, Normland; and (3) the waterworks of three mid-sized regional capitals in Ruritania, a fictional country located in sub-Saharan Africa.

The Key Issue

Anthropology usually enters the terrain of development cooperation within the scope of projects that are "close to the target group." These are measures that seek to establish as direct a collaboration as possible with those benefiting from the project and generally involve the transfer to farmers and breeders of information on, for instance, new cultivation methods, seed improvements, methods of livestock breeding, or loan systems. Impoverished city dwellers are sometimes integrated if the project deals with, for instance, preventive health measures, HIV/AIDS prevention and treatment, family planning, or environmental protection measures. Although all of these areas require some formal organization, the primary focus here is on altering cultural practices.

The lion's share of financial assistance in development cooperation, however, flows into the public infrastructure of developing countries. This includes formally organized systems—most of which were set up by the colonial powers—which aim primarily at securing education, health, communications, transportation, and administration. In most cases this involves large technological systems such as railroads, power supply systems, telecommunications networks, systems for drinking water, computer networks, systems of epidemiological prediction, and the like. The high priority allotted to this domain is evident in the list of measures that is generally drawn up after major political upheavals. Whenever a region is struggling for autonomy, such as Southern Sudan, leaders of the movement call for external support in a number of areas as soon as they sense the prospects of victory. Initially they try to secure the food supply. This requires that streets be made passable, bridges repaired, and markets created. Then they attempt to reestablish health care, which requires setting up a comprehensive organization, training staff, and obtaining medicine and equipment. Later they try to reopen the school system. In order to do all this, a system of administration, a tax agency, and judicial and police apparatuses also have to be reestablished. Development aid thus becomes a matter of governmentality.[9]

The leaders and their supporters stress that these measures will serve to maintain the newly gained self-determination. This claim is true to the extent that the new rulers mean political independence from a hated state apparatus, which they are in the process of overturning. They are not, however, referring to self-determination with regard to the rehabilitation of the basic infrastructure that was destroyed during the war of liberation. Potential outside supporters of this struggle for autonomy must first be convinced that the efforts are correct and proper according to their own expectations and norms. If the envisioned self-determination involves child labor, discrimination against women, intolerance of various religious faiths, or environmental destruction, then support will be difficult to find. For this reason, as a rule demands are adapted to the expectations of potential supporters, which triggers a problem of implementation at the local level. As soon as support gets underway, organizational structures and appropriate procedures must be set up to enable the transfer of funds, ideas, models, and artifacts. This poses a second and more fundamental problem. Insofar as the key internal issue requiring external support is the weakness of organizational structures and the unreliability of bureaucratic procedures, the entire process is caught in a vicious circle. The new, independent, and sovereign state apparatus has here reached the point at which the old oppressive state apparatus failed. Liberation movements generally emerge precisely in response to the nonfunctioning or perversion of what could be called the modern state apparatus and the institutions of civil society. Now, after the battle, an organizational structure has to be established *ex nihilo* for the importation of expertise and resources necessary to build up a functional and legitimate infrastructure. Needing an infrastructure in order to be able to establish an infrastructure, however, is a typical "Catch-22" situation.

From the perspective of anthropology, this problem can better be examined on a smaller scale and in calmer waters. One curious key term employed in the realm of development cooperation pertaining to the organizational forms enumerated above is "rehabilitation." The term refers to an organization or its technological system—such as a municipal drinking water system or a railroad company—which might still exist and have even been subject at one time to reform measures, but which has become so rundown after a period of time that it needs to be rehabilitated.

In practice, the issue looks something like this: A development bank reviews the application for a rehabilitation measure and then commissions a private contractor, a so-called consultant, to design the measure in detail and to implement it. After studying a mountain of documents regarding the state of affairs in the ailing organization as well as the most recent reform measure, the consultant generally comes to the conclusion that this past reform was not effective because it misconstrued the actual situation of the organization. Usually this is attributed to the fact that the previous consultant conceived the reform poorly. Consequently, the new consultant sets out to prepare a better plan for rehabilitation, but is confronted with the difficult problem of determining the actual situation within the ailing organization. This difficulty stems from the fact that the organization is unable to provide any reliable information, which was precisely the problem in the first place. Nevertheless, the consultant has to draw up a description of the situation and a plan for intervention as best as possible before the rehabilitation commences. A few years later it turns out that the organization has again failed to function properly and the next application for rehabilitation is submitted to a development bank. This is generally not even noticed because the organization has now been transformed into another organization, the subject of the current rehabilitation is worded somewhat differently, the application is submitted to a different development bank, different development methods have since become applicable, or simply because different people are responsible who are not familiar with the preceding events. Then a new consultant is sought because the previous one evidently did not do a very good job. This leads to the next round, which frequently proceeds according to the same pattern.

It is entirely possible that positive changes emerge as side effects of such rounds. However, in most cases these changes are not substantial enough for the game to be brought to a successful conclusion. One might even suspect that the continuation of this particular game prevents the emergence of another, presumably better game. If this is the case, then the most significant consequence of development cooperation is that it prevents better options from emerging. And this in turn gives rise to the next suspicion: that this obstruction in fact arises within the arena of development cooperation itself. In other words, it can be traced neither to mechanisms anchored exclusively in the society receiving the aid nor to mechanisms located solely in the so-called donor countries or donor organizations. The mechanism responsible for the fact that the game continues unchanged despite substantive doubts is in all likelihood connected to a problem of *representation*. It is precisely this assumption that I pursue in this book.

At this point attention is usually focused on the value of political programs and the legitimacy of political representation within the general context of the state order and within the specific context of development cooperation. Without questioning the significance and correctness of this kind of research, I have assumed in this study the perspective of the sociology of knowledge and, more precisely, have followed a science and technology studies approach. In doing so, I have concentrated on a different dimension of the problem than those usually selected in scholarly literature on development. Both development cooperation itself and the organizational structures it is supposed to set up aim to establish reliable technologies for remote sensing, monitoring, and control, which enable organized action from a distance that is independent of local loyalties and priorities. This is in essence an issue of representing reality through technologies of inscription and organizational procedures that have been detached from other subsystems of society in such a way that they cannot be subjected, for instance, to social, political, or economic criteria. The kind of thematic focus I have adopted here is anchored of course not only in the object of investigation itself but also in theoretical reflections. It is my hope that this study will contribute to three ongoing discussions, detailed in the following section.

The Ongoing Debates

(1) During the final third of the twentieth century, skepticism regarding issues of representation spread and became so radicalized, even beyond the discipline of philosophy, that a new situation emerged. The difference between a phenomenological-hermeneutic understanding of this situation and a newer (de)constructionist one can be easily identified in at least one respect. The Thomas theorem, made famous by Robert Merton, reads as follows: "If men define situations as real, they are real in their consequences." In accordance with this theorem, the social sciences and the humanities have been concerned less with the world itself than with what humans regard as the world. This distinction remains unproblematic as long as the Thomas theorem is not applied to scholarly definitions of reality. At some point between the mid-1960s and the mid-1980s, the inevitable self-application of the Thomas theorem finally occurred in the social sciences and the humanities, and quickly became one of the most prominent issues of contention within these academic disciplines.[10] Whatever insights might have been gained, a certain price has also been exacted for this shift in focus from the relationship between text and reality to the relationship between different texts, that is, the shift to discourse and intertextuality. Many of the newer metatexts are characterized by a lack of existential meaning, a fact that has disappointed numerous readers. Questions of "what" and "why" have been translated into questions of "how," without the authors of these works ever returning to the inevitable attempt to explain things.

For the limited purposes of my argument here, we can divide anthropological works into two camps. In one camp we detect sardonic elation among its members because they believe that others will soon arrive back at the point where they themselves have always been. Here positivists, phenomenologists, and hermeneuticists of the old school form an unholy alliance against "philosophical anxiety" and "too much" reflection. They contend that they have avoided the unnecessary detour of deconstructionism simply by remaining on the proper path (for positivists, that of Karl Popper, and for hermeneuticists, that of Hans-Georg Gadamer). The other camp insists that the perspectives arising from this detour have in fact provided essential insights into the nature of things. (Both "camps" are in fact highly heterogeneous and antagonistic assemblages; the only thing that ties the members of each camp together is the homogenizing ascriptions of the opposing camp). In the present study, however, I do not engage in a theoretical discussion about the consequences of deconstruction for anthropology. Instead, in my treatment of development cooperation I have sought to demonstrate how I think anthropology should approach the problem of representation after the writing culture debate.[11]

(2) With few exceptions, anthropologists have exhibited a stubborn avoidance behavior in regard to modernity, in particular to its *canonic institutions and citadels.* After returning from the tropics to the metropolis, anthropology—or whatever it should be called now that it no longer limits its scope to the periphery—regards itself as primarily responsible for the investigation of oral narratives and polycentric knowledge on the margins, in niches, behind the scenes, and in the underground. Mainstream anthropology has an affinity for marginalized life-worlds that perceive themselves as alternatives to the hegemonic assault of modernity, and it regards itself as an advocate of these tendencies. At the same time, anthropology fulfills this advocacy role from the perspective of written, scientific logocentric knowledge, which is itself the decisive trait of a modernity that hegemonically incorporates all forms of resistance. As a result, anthropology has a particular blind spot with respect to modern knowledge and the institutions that produce this knowledge.

It is in this blind spot, however, that everything that constitutes (post)modern society takes place: science, technology, law, and formal organization. This is where the daily struggle for access to and control over social development occurs through the creation of objective representations. The omission of these domains in anthropology means no more and no less than the claim that the constitution of worldviews—that which anthropology is responsible for investigating—does not occur here. The implication is that worldviews are something that other people have, people who don't know better, who still believe in honor and the nation, in devils, angels, and deities, in alternative medicine and flying saucers, people who believe in grand narratives and who adhere to diverse ideologies. Worldviews, according to this position, are illusions that always require the same kind of anthropological explanation: Worldviews correspond less to reality than to the social conditions of their constitution. In contrast, people who deal in or analyze financial markets, who fly around the Earth in satellites, who split atoms, analyze DNA chains, and clone plants and animals, but also people who engage in development cooperation and anthropology, apparently do not have (and do not construct) any kind of comparable worldviews. Anthropologists would otherwise feel compelled to investigate them. The implicit claim here is that such people see the world as it "really" is. In this way, the responsibility that anthropology assumes for "mistakes" that it

seeks to rehabilitate as antagonistic and yet equally valid worldviews results in an implicit—and therefore even more merciless—denunciation.[12]
(3) Insofar as modern transformational processes are closely tied to delocalization and translocality, the traditional localism of anthropology proves to be an impediment in understanding such processes. Although anthropologists as a rule seek to establish transitions between cultures, I have chosen instead to observe and investigate the translation practices of others. I focus on development experts as actors in interstitial spaces and on the boundary objects and traveling ideas of those actors:

The planning and implementation of development projects is a focus for massive cosmopolitan activity. In the night telexes chatter, linking clients in Kenya and Indonesia to consultants in California or the Cotswolds. Aid agency staff pick over policies on freeway and commuter lines bound for Washington and London. Contractors check their sums over breakfast. As the day closes on the other half of the globe, bureaucrats mark off the dusty minutes to their journey home, while the putative beneficiaries of these concerns cash the day's wages to buy maize or rice for the family meal. All are linked by the networks that projects weave. With a cast numbered and funded in billions this is one of the world's fastest growing, yet least analyzed, forms of collaborative behaviour.[13]

In keeping with the sense of this quotation, it would be misleading to suggest that this study deals with a Ruritanian development project or a Normesian development agency, with a consultant or an anthropologist in development cooperation. Although these actors do appear throughout the study, the book is not really about any of them. The focus is instead on the organization of what occurs between them, and this organization again raises the issue of representation. The players involved have to agree on representations that can be considered valid in all contexts.

At the same time, the focus on *interstitial spaces* raises the issue of agency. The development project chosen for this study has been examined as an archetypal case of *distributed agency*, in which several actors, dispersed over distant locations and social worlds and possessing differing webs of belief and conceptual schemas, have to collaborate in order to achieve something. On closer inspection, yet another category of agency—a nonhuman agent— is also involved. The practice of development is, as I attempt to demonstrate in this study, primarily a matter of selecting one of the existing globally circulating and highly esteemed models for development and adapting it to a local context. These models—for instance a particular form of commercial accounting or financing public infrastructures such as urban water systems—are attached for the most part to specific technologies. These models and technologies acquire an agency of their own precisely because they are disseminated and duplicated and in this process come to be endowed with an authority to define the best solution to a particular problem. At times the causality is even completely reversed, with traveling models searching for problems they might be able to solve.[14]

The Thesis

Let me begin with three exemplary propositions. Two of them are uttered on a daily basis in the arena of development cooperation. Proposition A reads: "Since the introduction of the structural adjustment program, things have been progressing in sub-Saharan Africa." Proposition B reads: "Since the introduction of the structural adjustment program, things have been getting worse in sub-Saharan Africa." Both propositions relate to the same ontological order and agree that denotative propositions of this type are possible and can be verified against reality. Propositions A and B only disagree on the particular facts, not on the possibility or accessibility of these facts—which are taken in this case to be indicators of prosperity and poverty.

A third type of proposition is uttered less frequently and appears to be radically different. Proposition C reads: "You can only affirm proposition A or B if you trust the procedures that generate the required indicators; however, there is no good reason for such trust because these procedures cannot overcome their own indeterminacy and are inherently interested." Although proposition C clearly has a different logical status—it refers to sentences, theories, and methods rather than states of affairs in the "real" world—it nevertheless shares a common epistemological problem with the other two propositions. In order to refute Propositions A and B, proposition C necessarily relies on similar ontological presuppositions in order to assert the inadequacy and indeterminacy of "development indicators." Furthermore, it must presuppose at the very least the existence of these theoretical entities and claim itself to represent them adequately. In other words, objectivism (A and B) and antiobjectivism (C) belong to the same class of propositions; they merely relate to different levels of reality. Their circular relationship makes apparent the paradox of providing an ultimate grounding for propositions that claim to be true.

Hence proposition C does not resolve this paradox, although it may provide an important addition to the dispute between proposition A and B by introducing a dimension of reflexivity. Nevertheless, if taking a position is a necessary and unavoidable prerequisite for making a decision and assuming liability for that decision, then objectivism is ultimately an indispensable rhetoric. It is something we should seek to improve upon, even if it ultimately remains an unattainable horizon for humans as finite beings. This kind of self-reflective objectivism also protects political systems from ideological blindness, insofar as they are neither able to invoke an ultimate truth—whether religious or ostensibly scientific—which may prove to be erroneous in twenty years, nor are they condemned to hold all positions to be equally plausible. If this is the case, then the following is also true: Before one can claim that a proposition is correct, one must first concede as a condition of possibility that propositions never simply depict realities but always already order that reality conceptually (to borrow from Max Weber). According to this position, every representation inevitably has its own blind spot: It is specifically situated and cannot stand outside its own position and its own distinctions and is for this reason unable to ground itself independently. Although blind spots can never be entirely eliminated, they can nevertheless be "repositioned" as needed (as Niklas Luhmann would put it). "Need" in this sense is a pragmatically evident purpose that cannot be simultaneously reflected upon. "To reposition" means to seek out an observation point with a blind spot that can be tolerated for the time being. If this paradoxical repositioning succeeds, the newly attained perspective allows existing possibilities to be unblocked—but this always occurs at the expense of blocking other possibilities in the process.[15]

The consequences of proposition C for propositions A and B are as follows: Whether one believes that things are "progressing" or "regressing" in sub-Saharan Africa is certainly largely dependent on the metanarrative that one subscribes to, especially given the fact that without such a narrative it would be extremely difficult to distinguish between progress and regress at all. One also requires technologies of representation and calculation in order to prove whether it is A or B, but one cannot at the same time prevent these technologies from exerting a performative effect on the object of enquiry. This deconstructive conclusion, however, leaves untouched questions of meaning and existence. In terms of practical life, humans have to try to resolve the point of contention between propositions A and B and to seek better solutions for moral, juridical, and political reasons. In doing so, they must continuously distinguish between correct and incorrect propositions in order to select the proper path. They will have to do this even when the doubt implicit in proposition C has expanded into a general suspicion toward all forms of objectivism. However—and this is what counts—they will be able to do this successfully only if they bear in mind that they are necessarily standing on thin ice here.

According to the thesis of this book, the epistemic community of development experts generally believes that they are standing on firm ground. They attempt to drive their own constructions of reality into this ground like tent stakes in order to securely bind development projects to them. Because they are actually operating on thin ice, however, their actions give rise to problematic consequences. Beyond this critique, I also attempt to demonstrate that the objectivism of development discourse should not be traced back to Western universalism and its hegemonic claims. It is instead the transcultural processes of negotiation and decision making that produce objectivistic definitions of reality.

From this vantage point, the only remaining way to avoid getting hopelessly entangled in inappropriate definitions of reality would be (borrowing from Luhmann) "to shift the observer perspective" (or "to reposition the paradox"), what Goffman would call a "frame change": What we are compelled to regard as real within the framework of a particular negotiating or decision-making situation can, within the framework of reflecting on that situation, be traced back to the conditions set by the initial framework.[16] For the purpose of my argument and in particular for my ethnographic presentation, the term *code switch* seems more appropriate. In phenomenological terms, only code uses and code switches can be observed, whereas frameworks and perspectives remain the explanatory models on which they are based. In order even to participate in a mutual game, players in an arena must agree on a universal code that appears to be comprehensible in all frames of reference. I will call this a *metacode*. The same players, however, shift facilely to a *cultural code* when they comment on the moves of other players before and after the game and attribute these moves to other players' models of cultural orientation. The arena of development cooperation is characterized by a precarious situational alternation between metacode and cultural code.

The Vocabulary

If the validity of a representation cannot be determined in terms of correspondence theory—even if fidelity to reality remains an indispensable criterion—then the conception of representations needs to be defined differently. Representations are elaborately fabricated in both development discourse and anthropological discourse. To the extent that they are implemented and thus institutionalized as valid versions of the world, these representations play a constitutive role in defining the world and the practices thereby legitimated. In order to understand the phenomenon called development, it is therefore essential to investigate the transitions between the representations and the practices with respect to how they are institutionalized and deinstitutionalized.

To represent means "to imagine," "to depict," "to act as a proxy," and "to bring to mind," but also "to provide an impressive example of something" (as in "representative architecture"), as well as "to be typical" in the sense of being "statistically representative." In all of these meanings the same figure shines through: In place of an absent or unattainable reality, a surrogate, a copy, or an advocate is presented. This inevitably raises the question of the proxy's authorization (i.e., its legitimacy or validity). Does the advocate of the speechless correctly represent those who do not have a voice? Does that which is present adequately depict that which is absent? How can distortion and deception be avoided in the process? This question points to the basic paradox that easily leaves the initially undaunted observer petrified, as Luhmann has ironically noted: In order to become reality, reality has to be objectivized in a representation. Conversely, however, the objectivity of a representation is never exhausted in the represented reality. What role then do representations play in the construction of represented reality? What role do fictions play as depictions that do not represent any reality? And what role do undepicted realities play?

Only when observers are dealing with something directly observable ("she turned on the faucet and water came out"), that is, when they share a common frame of reference and stand in direct communication with each other, can they agree immediately about its reality. However, in those cases in which people communicate over distances (both spatial and temporal) and in which the objects in question are complex, the matter is no longer so simple. Because individuals do not have the external referents directly in front of them, they become aware of the significance of representational practices. Representational practices produce transitions between realities and depictions of those realities.

However, it is not only because of the spatiotemporal distance from an external referent or the complexity of that referent that a direct test of the correspondence between representation and reality necessarily fails. Another decisive factor is that something regarded as an external referent in one frame of reference can be regarded as a socioculturally conditioned construct in a different frame of reference. Nevertheless, it is sometimes necessary for people to bridge these gaps in order to reach an understanding and act jointly. A typical example of this is development cooperation. When the representation of something can be transferred from one frame of reference to another without that representation losing its validity—although the validity might and indeed usually does change—we can speak of the corresponding representational practice as *translating*.

Translation occurs when an idea or a thing is carried over from one idiom to another, from one culture to another; or when an idea or a thing is connected to another idea or thing in such a way that its effect is intensified as a result (as is the case, for example, with a pulley system or a bicycle chain); or when an idea is manifested in a practice or a thing and vice versa. All of these different meanings have a common denominator: Translation brings together things that are separate; it establishes a relation and mediates between multiple elements and makes them compatible and comparable. In this sense, translation also produces commensurability by establishing gauges and metacodes. A form emerges that did not previously exist. The act of translating generally raises the following question: Is this an accurate translation or has the meaning been altered?

This generalized suspicion is not entirely unjustified since translation is a procedure necessarily tied to the power that emerges in the process of representation and thus opens up diverse potential for manipulation. Viewed in this light, political representation is itself a process of translation: In order for many voices to become one, they must be translated, and, not surprisingly, politicians speaking in the name of their constituency are habitually suspected of manipulation. It is not possible, however, to avoid this manipulation entirely. In order for ideas (usually inscribed into models and artifacts) to circulate from one social world to another, from one frame of reference to another, they must be adopted, appropriated, and altered. Ideas are evidently unable to go very far on their initial impulses, with only the energy from their original frame of reference. To be transferred they have to be transformed, that is, translated. Every act of translation is inevitably also an act of performative omission and addition; otherwise the translation chain would break. Every act of translation is thus also an act of creation, producing something that did not previously exist.

An important example of translation in this sense is the construction of representations in order to grasp a complex issue. Observers who want a bigger picture of reality than they can see with their own eyes and from their own particular vantage point are compelled to construct a series of mediations and proxies between themselves and reality. The first step is to decide which tangible substitutes of the whole should be gathered. One person might begin by selecting an interesting location; another might go to a library or an archive; and a third might seek out and interview relevant experts. No matter how this selecting and gathering occurs, the next step will invariably consist of viewing and ordering the various substitutes. The selected and classified representations are ultimately combined into a bigger picture, which was the reason for engaging in this laborious process in the first place.

But what happens when skeptics come along and contest the validity of this picture (which will inevitably occur at some point)? Specific to this bigger picture is that there is no single vantage point from which all of the external referents can be seen. It was precisely for this reason that the picture, as a representation of something whose actual existence remains debatable, was introduced. Has anyone really ever seen "society," "the economy," "justice," "power," "progress," or even "development"? Thus it is impossible to take this bigger picture or text and measure it against "development" to verify the correspondence between reality and representation. Skeptics would instead have to retrace the steps back along the entire path. And in doing so, they would inevitably discover that the individual substitutes that were used to compose the bigger picture were only substitutes for other substitutes. Skeptics will move from one document to another, and when they finally believe they see the light at the end of the tunnel, they will encounter substitutes there as well. Perhaps they will ultimately reach a point at which nothing is concealed, but the exposed reality will offer no immediate answer to the larger question. The skeptics will discover not "development," but at best a tangible artifact, in our case, for instance, a new water meter.

For this reason, the objectivity of a representation cannot be resolved according to the model *adaequatio rei et intellectus*, the equation of thought and thing. The issue is instead the clarity and the methodological validity of the aggregation used to compose the bigger picture from the individual pieces. Because these individual pieces are not direct substitutes for an external reality but instead bring forth a cascade of further substitutes, one is never dealing with a single referent but rather with a diversity of internal or *transversal referents* that have been organized into a chain such that they support themselves as they proceed along it. From this perspective, a representation is always a cascade of re-re-...representations. Because the practice of representation is best understood as a translation, I will call this a *translation chain*.

The thematic focus and the structure of the present study are based on the assumption that the production of transversal referents is the key technique for both development cooperation and anthropology. When I say that I am writing about the making of development, I am therefore addressing precisely this point. Hence this study will report not about what actually happens "on the ground" (for instance, in the waterworks of Baridi, Mlimani, and Jamala), but rather about what actually happens in the interstices and the translation chains. In examining this problematic, I focus on technologies of inscription and representation.

Empirical Considerations

(1) The locus classicus for the emergence of *methodological agnosticism* in anthropology has been the analysis of magic, witchcraft, and the belief in all kinds of spirits. While locals usually make the substantialist claim that their belief in spirits is caused by the evident existence of spirits, incredulous anthropologists are compelled to seek more circuitous answers because this evidence does not make much sense to them. As long as evidence of the existence of spirits remains the issue, few inhabitants of the US–European world would regard it either as mistaken or presumptuous to replace the answers offered by believers with what they consider to be more plausible ones. They usually find it more convincing to derive the belief in spirits from the social order in which it arises. This is because in their view it simply makes more sense to assume that there is no such thing as spirits. As I will argue throughout the present study, I think that this skepticism toward propositions based on occult, irreproducible claims makes sense primarily for legal and moral reasons rather than epistemological ones.

With more familiar representations, however, the matter appears in a rather different light. If, for example, an anthropologist were to set out to investigate why some people believe that the Earth is round, most inhabitants of the US–European world would regard this as a bizarre enterprise because they know that the Earth is in fact round. Nevertheless, if anthropology is concerned with questions about how worldviews arise and how they change, even in cases such as the shape of the Earth, it can only fulfill this responsibility adequately if it proceeds consistently and elevates the agnosticism that it brought home from the tropics into a methodology. This proves to be a particularly fruitful approach wherever scientific knowledge is produced. Competing bodies of knowledge develop for all controversial topics, which means that decisions about such issues are ultimately made under conditions of uncertain expertise. For this reason, the key question for an anthropology of science, technology, law, and organizations should be just as agnostic as the key question for an anthropology of occult practices: How can people know what they believe they know?

(2) In the domain of development cooperation, anthropologists deal with locals who have the same level of education, who often earn considerably more money than they do, and some of whom occupy high-ranking positions. These people are able to effectively defend their terrain against unwelcome intrusions. They regard anthropologists not as representatives of a superior culture, but as members of the insignificant genus "social scientists." For anthropologists, this situation is called *studying up*. The important point here is that if anthropologists follow the principle of methodological agnosticism, their view of the locals inevitably annoys the latter. While locals—in this case, development experts—labor to increase their certainties in order to be able to act responsibly, the anthropologist hovers around them, peering over their shoulders with an interested but skeptical eye. The underlying assumption is "You may be the expert here, but I see something that you cannot see, and that is the way in which your ideas are dependent on your frame of reference." In the context of studying up, this annoyance can become acute at any moment and result in the anthropologist's exclusion from the terrain of study. In order to overcome this seemingly insurmountable barrier I opted for true participant observation and worked as consultant for several years. In doing so I shared with the other players the same responsibilities and the same risks of making mistakes and of losing face and money as well. This meant that I had the same access to information that other development players had. It also meant that I did not have the privilege of an outsider, whom insiders sometimes entrust with confidential knowledge that they would never share with each other. However, it is less difficult to overcome this internal barrier than to gain access to the field—at least if you happen to be an anthropology professor and have enough time on your hands.[17]

The Textual Strategy

The genre of the ethnographic report generally operates according to two authorities or perspectives. There is, on the one hand, the authority of the locals concerned with their own "native point of view" (for example, the Trobriand Islanders). Usually, they appear in the ethnographic report as integrated believers who see the world as it should be seen according to their culture's notions of reality. On the other hand, there is the first-person narrator who plays the role of both observer and skeptic (in the case of the Trobriand Islands, Bronislaw Malinowski). This narrator wanders through the mental topography of the local environment and reports his impressions to readers: "I was there, I saw it myself, and consequently I am authorized to report truthfully about the things that exist there."

Literary narratives, in contrast, generally operate according to three authorities. A first-person narrator represents the empirical author in the text, playing the unobserved observer of events described in the narrative (the first authority in *Madame Bovary*, for example, is Flaubert). This observer follows a skeptic who traverses the mental topography of an epoch or a milieu (in the selected novel the figure of Emma Bovary is the second authority). By following the skeptic, the first-person narrator encounters representatives of normality (the third authority, Charles Bovary and many other figures of the novel), who can now be observed through the lens of the skeptic. Because the skeptic often scandalizes and violates the accepted norms and conditions of society, the observer renders visible to readers the resilience and internal logic of those relations (and thereby depicts the *moeurs de province* or "provincial mores," the subtitle of Flaubert's novel). In addition, the text can be refined self-referentially through the introduction of a fourth authority: The empirical author (the first authority, for example, Vladimir Nabokov) delegates the role of the first-person narrator and the author in the novel *Pnin* to a figure located in the text (the second authority, N.), while the rest of the triangular constellation remains unchanged. This directs the reader's attention to the fact that the text is fabricated. The internal rules of fabrication become observable, thereby thematizing the reciprocally constitutive transitions between representations and reality.[18]

Because I am concerned in this study with representational practices that I not only describe but also perform in my writing, I have constructed a narrative with four voices. As the empirical author, I am inevitably the first voice. However, following this introduction I assume the role of the first-person narrator again only in the fourth part of the book entitled "Trying Again" (except for the endnotes, which throughout the book are my own authorial insertions). The second voice in the narrative, a fictional anthropologist named Edward B. Drotlevski—who is my first replacement in the text—guides readers through the sites and the mental topographies of the locals in question. Drotlevski appears as the author of the first three parts of the book entitled "Belief," "Doubt," and "Searching." On his odyssey through the world of development cooperation, Drotlevski encounters locals who mistrust the representations of their world (the third authority) as well as those who trust in them (the fourth authority). As the putative author in the text, Drotlevski lets the latter speak for themselves in part one ("Belief"), while the former are given the same opportunity in part two ("Doubt"). In this narrative, doubt is represented by science in the character of the anthropologist Samuel A. Martonosi—my second *locum tenens* in the text—whereas belief is embodied by diverse figures who make decisions, assume responsibility for the development cooperation, and are paid accordingly. The most important protagonist on the side of belief is the consultant Julius C. Shilling.

Instead of imitating—or borrowing—the voices of the Tanzanian (alias Ruritanian) managers of the water utilities, as classical ethnography would have done, I invented the anthropologist Martonosi, who observes the local actors and practices and is himself observed by another textual figure, the second anthropologist Drotlevski. In doing so I experiment with the core business of anthropology—that of "giving a voice to the voiceless"—in a way that is intended to demonstrate to readers the pitfalls of such an endeavor.[19] At the same time, this textual strategy seemed to me appropriate for two additional reasons.

First, the local managers do their best to avoid explicitly articulating their own views and interests. They tend instead to inconspicuously submit themselves to the globally circulating models that have the aura of cutting-edge solutions drawn from the most technologically advanced countries of the North. This is not the whole story, and there are certainly other things that occur behind the scenes. The crux, however, is that precisely these things are never directly represented in the official interactions and negotiations of development cooperation, and it was my main intention to examine how development functions at this level. It is usually assumed that this willing submission is the only possible tactic in the face of the overwhelmingly powerful and salient strategies of Western development experts.[20] While my investigation does in principle confirm this assumption, it also indicates that there is significantly more room for maneuvering even at this level. In other words, I believe that the hegemonic dominance of the Western experts and Western models do not provide a sufficient explanation for the tactics of the Southern partners. However, because this is virtually impossible to prove empirically, I have concentrated instead on examining how the tactics of Southern actors contribute to the incontestability of Western models.

Second, and perhaps a bit provocatively, I wanted to insist that the voices missing in the social science literature on development and technology are not necessarily the "native" ones. It is, on the contrary, the voices of technical experts from the North—such as that of Shilling—that are rarely heard. Although we do know a lot about development policies and the people who vociferously present them to the public, the technical experts and their complex organizational arrangements, procedures, and technologies remain in the shadows. This is not unlike many other fields: Architects, for instance, are visible whereas engineers remain anonymous; musicians are celebrated stars while the makers of their instruments and the engineers of their recording studios are unknown. It is this problem of voicelessness in the interstitial spaces that I wanted to examine in the case of technologies in development cooperation.[21]

In order for a story to run its course, it needs a plot.[22] As preparation for a research trip, Drotlevski conducts several interviews. In chapter 1, von Moltke, as the representative of the development bank, explains the significance and purpose of this particular development project. In chapter 2, Shilling, the consultant, describes its practical implementation. In chapter 3, Martonosi, the organizational anthropologist, explains why things cannot proceed as intended. In chapters 4 and 5, Drotlevski then reassumes the reins. After concluding his preparations in Europe, he travels to Ruritania to inspect the project and form his own view of the conflict between belief and doubt. Finally, in chapter 6, I bring together all the reports and review what actually takes place in the interstices.

BELIEF

1   The Solid Ground of Development Policy

Preliminary Remarks : Edward B. Drotlevski

The architecture of the lobby at the Normesian Development Bank (NDB) led me straight to a bulletproof-glass reception booth. Without a word a form was pushed through the narrow slot of the receptionist's window and, as instructed, I wrote down my name, my institutional affiliation, and the person at the NDB with whom I wished to speak. The receptionist announced me over the telephone as Professor Edward Drotlevski from the University of Urbania and then indicated that I should be seated in the Bauhaus-style black leather suite across from the booth.

A triptych on the wall high above the leather chairs caught my eye. It depicted an African landscape with rhinoceroses radiating an irrepressible vitality. As the rooms in this building were dedicated to reflection about Africa, I found myself asking the following question: How do these wild African rhinoceroses correlate to the modern disciplining of the African continent? Later I learned that there had been an exhibition of five young female artists entitled "Investigations: Women Paint Africa" in the NDB building between September 9 and October 11, 1992. Several of these paintings, including the triptych, now decorated the rooms of the bank. At the exhibition opening, a high-ranking bank representative had declared that the bank, like the artists, was engaging in investigations of Africa. In this regard, he continued, the bank had distinguished itself from colonial rulers, who had attempted to impose a foreign order on Africa. As I contemplated the massive painting, I thought about the different alternatives that might have been used to decorate the lobby—aerial photographs of a colossal dam, for example, or a new and shiny water processing plant? No, in 1997 it was no longer appropriate for the NDB to drape itself in such images.

The director of the sub-Saharan Africa division, Dr. Johannes von Moltke, was sitting in a dauntingly spacious room with a long window front that extended all the way down to the luxurious carpet and faced out onto a quiet side street. The exclusive office was virtually empty. Instead of shelves filled with files and books, the walls were covered with abstract paintings. We sat down at the large conference table, and I was allowed to turn on my Sony tape recorder. The conversation gathered momentum quickly and easily. Despite the fact that it lasted more than two hours, we were not interrupted by telephone calls. The fact that von Moltke had his secretary hold all incoming calls was an obliging gesture of recognition for his interlocutor. Von Moltke's own remarks were so clear and systematic that I have decided to let him speak for himself here.

NARRATOR : JOHANNES VON MOLTKE
Location : Urbania, Normland
Date : July 14, 1997

The Project within the Framework of Development Policy

The chief business of the Normesian Development Bank (NDB) is so-called financial cooperation (FC). The objective of FC is to facilitate access to the *production potential* of developing countries by making capital available. This usually means rehabilitating and modernizing preexistent facilities and infrastructures, which is presumed to be possible given the necessary capital. As a state bank, however, the NDB must also avoid taking market shares away from commercial banks, since that would undermine its own goal of promoting private-sector development. Our loans are so-called soft loans, which can assume different forms depending on the country and the respective development measure. In some cases there is a ten-year grace period followed by forty years of repayments at an interest rate of 0.75 percent. Countries classified as "least developed" receive grants only. According to a Normesian distinction, in addition to FC there is also so-called technical cooperation (TC). TC aims to raise the *performance of human resources and organizations*. The Agency for Overseas Development (AOD), which is far better known both domestically and abroad than we are, is responsible for TC.

The water supply project in Ruritania is conceived as an accompanying measure of a larger investment. It has a budget of three million dollars and is financed from a special fund of the Ministry for Development Cooperation (MDC). This special fund serves to support basic and further training worldwide and is utilized as an aid grant so that neither interest nor amortization payments accrue. The aims of the accompanying measure correspond to the MDC's present conception of development policy. This conception is based on the following insight, which is now generally recognized worldwide in light of the failure of dirigiste socialist models: A market-oriented socioeconomic order and a social system geared toward citizen participation in the political process offer the best prerequisites for humane development.

The drinking water supply project for the three regional capitals—Baridi, Mlimani, and Jamala—has been conceived precisely with this in mind. First of all, the project is designed to decouple the waterworks from the Ruritanian regional administration and bring it closer to the city residents through a new organizational structure. Second, the project aims at subjecting the production of water as a public commodity to a regulatory regime that will facilitate commercial operation. The project continues to follow the model of globally sustainable development through productive economic growth, social justice, and environmental sustainability.

The *economic growth* of a city is fundamentally dependent on a well-functioning infrastructure. This requires, first and foremost, a reliable and affordable supply of high quality water for private households as well as for commercial companies and industries, not to mention hospitals, schools, and public institutions of all kinds. But sustainable and sound economic growth also requires that a waterworks can be permanently maintained economically and technically, and if necessary even expanded, through its own revenue. Above all, stable economic growth means avoiding dirigiste distortions that would undermine market economy regulation methods and lead to an economic disaster, as Ruritania has recently experienced. At the end of Ruritanian socialism water was virtually free of charge, but it hardly ever flowed.

With respect to the development policy goal this means in concrete terms that the price of water has to correspond to the actual production costs that accrue in a certain area. This is the only way to guarantee that a city's growth rate remains reasonable relative to the natural, technological, and economic options for the water supply. What should be avoided are subsidized water prices that allow cities whose water system production costs are relatively high to expand at the same pace as cities with lower production costs. Especially when industry is brought in, the price of water should have a regulating effect. Since a municipal waterworks is a natural monopoly, it is also important to guarantee that only the really necessary expenses, but not the organization's inefficiency, contribute to the production costs. In the absence of real market mechanisms, intelligent regulatory mechanisms that effectively monitor the economic efficiency of the monopolist must be introduced to this end. That is the goal of our project.

The project will also effectively promote a second major aim, *social justice*. As long as the charges for water remain below the actual water price—that is, the production costs including the facility write-off—the arrangement will require that ensuing uncovered costs be paid by other state resources. However, since the main source of Ruritanian tax revenue is agriculture, the old, low, and therefore socially acceptable water rates in Ruritanian cities could more accurately be referred to as transfer fraud. If subsidies based on socio-political considerations continue to be necessary in the future, they must be justified on both economic and political grounds. This presupposes the correct assessment of the necessary production costs from an economic and technical standpoint as well as transparent organizational structures that allow us to determine where specific subsidies went and why, and what has to be considered in the future in order to eliminate the need for them.

The project also contributes to greater social justice in another respect. Since 1991, when tap water was de facto no longer supplied to Ruritanian cities free of charge, debate about the payment morale of customers and the payment collection rate of the waterworks has become increasingly heated. According to the data available, less than half of all customers pay their bills at all. That means that nonpaying customers indirectly exploit paying customers. For this reason the project is seeking to raise the collection efficiency of the waterworks in Baridi, Mlimani, and Jamala to ninety percent.

The long-term economic rehabilitation of the waterworks will ultimately enable them to appropriately renovate and possibly even expand their pipe system and their production capacities. The ability to make replacement investments from their own revenue means that the present generation will not operate the facility at the expense of future generations. Self-financed expansions would benefit especially those districts of the city whose infra-structures are neglected and which are usually inhabited by people from the poorest strata. A locally oriented, commercial operations model would encourage city residents to acknowledge the waterworks as their own property worthy of protection and therefore to treat it with due care.

The third major aim of Normesian development cooperation directly addressed by the project is *environmental sustainability*. The most effective protection against wasting water is to charge money for it, to collect payments consistently, and to combat illegal taps. The planned commercialization will also put the waterworks in a position to pay the responsible authorities the groundwater fees that are now due in Ruritania. The River Basin Authority was established with European support in order to develop general proposals for conserving water in the area and facilitating a meaningful balance between various water uses (energy conversion through hydroelectric power plants, agriculture, animal husbandry, and drinking water).

The History of the Project

In the 1970s and 1980s, the municipal waterworks were part of the regional governments and were affiliated at the same time with the water ministry in the capital Baharini. Even if this administrative structure did not make any sense, it did not seem appropriate at the time to address this issue within the framework of the FC project. Nor did it appear to be the right time to broach the issue of project execution in any fundamental way. It was simply accepted that the central water ministry was the project-executing agency and thus bore responsibility for construction measures in the waterworks of the regional capitals. In any case it was assumed that the issue would be approached from technical and public health perspectives and that economic and administrative policy perspectives would be factored out as much as possible. Everyone was aware at the time that none of the Ruritanian waterworks produced sufficient revenues to cover their own operating costs, but this was not an issue in development cooperation in the early 1970s.

An application by the Ruritanian government in 1969 initiated the first project, "Jamala Water Supply." After the usual inspection process with numerous modifications, the first loan agreement for 6 million dollars was reached in 1973. Cholera epidemics broke out in Jamala in the early 1970s as a result of catastrophic conditions in water supply and sewage treatment. Thus the project undoubtedly served a concrete, legitimate goal. By fall 1976, project funds had been increased to 24 million dollars. From an engineering standpoint the project addressed the root of the problem: In order to improve water supply, first there has to be water. For this reason the project focused on the groundwater extraction system. The existing system was ancient and ramshackle—it dated from the colonial era—and was located right in the middle of Jamala due to the urban expansion. As a result, the groundwater was contaminated by cesspools and the broken-down sewage system. Within the scope of the project a reservoir, a pumping unit, and a water processing plant were built outside of the city, and a water pipeline was connected to the distribution tanks located in the hills on the outskirts of the city.

When the new water extraction and conveyance plants went into operation in Jamala in December 1978, it became evident how inadequate the distribution system was. Although the system could now be filled, a rather considerable percentage of the water was lost in the network. In response the NDB commissioned a study on the condition of the distribution network and then initiated a rehabilitation program. Through a series of complications that would be difficult to reconstruct today and can be traced in part to the fact that the water ministry served as the project-executing agency, the corrective repair measures were repeatedly postponed so that the distribution network was not repaired until 1987, ten years after the opening of the water treatment plant. Owing to the relatively minor scope of the project only the most blatant defects could be corrected. An unacceptably large share of drinking water continued to be lost through leakage, and water pressure remained too low for satisfactory service. In the meantime, the first rehabilitation measures of the water treatment plant and the pump stations were already on the agenda. The investment measures now encountered the notorious problems: improper usage, irregular and improper maintenance, lack of replacement parts due to shortage of funds, and inadequate organization. At the beginning of the project studies around 1970, there were already suspicions that not only the distribution system was full of leaks, but that payment collection was itself replete with accounting holes. In 1978, this deficiency was identified explicitly for the first time as an obstacle to the success of the project. It was expected that Ruritanian officials would deal with this impediment, but this did not occur.

The second waterworks connected with the project was located in Baridi. The background here is similar to that in Jamala. Within the scope of financial cooperation, the existing water supply in Baridi was overhauled between 1975 and 1977. This plant, which is still in operation today, transports spring water from a nearby mountain into the city through the force of gravity. The plant was repaired again between 1981 and 1985. At that same time, there was exploratory drilling to investigate possibilities for expanding water supply through groundwater. "Baridi Water Supply," the main project, was finally initiated in 1982. As part of this project, an extensive field of wells below the old springs was established. In this way it was possible to develop a sound groundwater supply for Baridi to compensate for the seasonal fluctuations of the spring water, which also continued to be used. Along with diverse smaller accompanying projects, around 20 million dollars was invested in Baridi's water supply system. Overall, investments in the water-processing plants of Baridi and Jamala have amounted to more than 60 million dollars.

In Baridi, these construction measures could only begin in 1985 because of a series of technical problems and organizational confusion. As a result of further unexpected difficulties, the new plant was not opened until 1990. As in Jamala, after the plant was opened it was discovered that the pipe system had unacceptably high water losses and the accounting system was also full of leaks.

Initially, however, there were more pressing problems in Baridi. In comparison to the old system's use of gravity, the running of the well field proved to be quite maintenance intensive and prone to failure. Power outages and voltage fluctuations led to frequent damage to the electrical underwater pumps. Finally, there was not one specialist in Baridi or even in the entire country who was able to rewind copper wire on the electric motors of the pumps. At first they had to be sent to Normland by airfreight for repairs. Later, a company was located in a neighboring country. The NDB then established a supplemental program to improve operations and a training program for electricians. People learned in the meantime to live with the voltage fluctuations and blackouts and to repair the burned-out pumps themselves.

It was ultimately discovered that while electricity prices in Baridi had risen dramatically in the time between the planning phase and the start of operations in 1990, water prices had hardly risen at all. Before the project, the waterworks had had almost no electricity costs whatsoever; now approximately seventy percent of the waterworks expenses were for electricity for the modern pumps. As in Jamala, this gave rise to the following problem: How can the waterworks equipped with new facilities survive on their own both economically and technologically? In other words, how can we ensure that the two waterworks neither deteriorate nor remain permanently dependent on development aid?

The technical improvements at the waterworks in Baridi and Jamala had been implemented entirely in terms of increasing production potential, as officially defined by financial cooperation. Given these considerations, we gradually realized that they would be condemned to failure if the project was not sensibly integrated into larger organizational changes. The interaction between two developments contributed to our insight here. On the one hand, in connection with the start of operations in Jamala in 1978 and Baridi in 1990, the financier realized that simply owing to lack of funds the investments could not be maintained and operated as stipulated by regulations.

The desired life expectancy of the plants would be dramatically reduced without improvements in the realm of bill collecting and thus in the entire realm of organizational efficiency.

On the other hand, in the twenty-two years between 1969 (when the Ruritanian government applied for the waterworks project in Jamala) and 1991 (when the NDB commissioned progress assessment inspections in Jamala and Baridi) there were several changes that affected virtually all dimensions of development cooperation with Ruritania. This period included the brief blossoming of Ruritanian socialism as well as its doleful self-dissolution. In connection with this latter development, the idea that a centralized state could plan social development lost dramatically in credibility throughout the world and was symbolically laid to rest with the fall of the Berlin Wall in 1989. With this historical break, the competition between the Communist and the Western worlds for influence over the nations of the southern hemisphere became a thing of the past, a development that considerably increased the negotiating strength of donor organizations.

But even before this, in the early 1980s, after three decades of development policy, economic and political difficulties were apparent in the poor nations of the South in the form of stagnating and sinking per-capita incomes, balance of payments and debt problems, scandalous income differences, social tensions, and serious supply problems. These facts could no longer be overlooked, and people gradually dared to point out that the problems here could in part be traced back to self-made errors in economic and social policy. Politicians responsible for development policy and the relevant government development institutions could simply no longer afford to continue with business as usual at home, where they drew their monies from tax revenues. They could no longer act as if their projects—for example, the waterworks projects in Baridi and Jamala—would achieve sufficiently positive results as ideal microworlds situated within problematic macroenvironments. It had become all too clear that the mechanisms functioned more the other way around. The mass media had already begun to focus primarily on blunders of this nature.

In light of these issues, the concept of structural adjustment was developed under the auspices of the World Bank (WB) and the International Monetary Fund (IMF) in the 1980s. The three central points to structural adjustment are: (1) dismantling state intervention into markets; (2) reducing the state and administrative apparatus to a necessary minimum; and (3) reestablishing an investment-friendly climate by paying attention to the balance of mobilizable resources to consumed and invested resources. In practical terms, this means supplementing individual projects with large-scale reform packages that transform the economic and administrative-policy structure of the country concerned. This means massive interventions in the domestic affairs of the countries of the South according to the slogan, "It is impossible to live beyond your means over the long run." Within the framework of structural adjustment the elites of these countries were addressed as the parties responsible for executing the reform packages if they wanted to benefit from development cooperation. In general terms, loans were no longer distributed as reparations but rather in exchange for the fulfillment of certain conditions. In the language of the World Bank, "investment lending" was replaced by "policy-based lending." The overall economic successes of structural adjustment policies in the countries south of the Sahara are obvious. Between 1994 and 1996, positive economic growth has been measured in eighteen reform countries of sub-Saharan Africa.[1]

The Goals of the Project

Looking at our investments in Jamala and Baridi, we had to admit that over the course of the years the collection efficiency of the waterworks remained consistently low. However, even if this weakness had been resolved, the low Ruritanian water rates would have meant that potential revenues remained insufficient for proper operations. In addition, revenues still did not remain in the works themselves, but rather had to be paid to the ministry of finance through the regional governments. Finally, the budgets for the waterworks in Jamala and Baridi approved by the central government were consistently lower than the (already modest) sums that the works themselves were able to earn. Especially this last point meant that although NDB investments did in fact generate revenue, this was then used in part to subsidize the regional budget in other areas. As a result, Ruritanian engineers had to economize wherever possible without directly reducing the water supply. As is often the case in such scenarios, the waterworks reduced their expenditures for maintenance to the bare minimum and scaled down the operation standards below acceptable levels. Because of this, we canceled any further engagement in the water sector in 1991. This was our justification for rejecting an application by the Ruritanian government for new investment in the waterworks of the regional capital Mlimani based on the model of Baridi and Jamala.

In search of an alternative, we commissioned a sector study in 1992. This study identified the organizational environment as the reason why our projects did not have a positive effect. The administrative policy of the entire sector was organized in such a way that any initiative or desire to reform was undermined by the fact that people charged with responsibilities and decision-making powers lacked sufficient motivation and the necessary competence. Either responsibilities tended to wander up to higher levels in the administrative hierarchy or new authorities were created in the course of reforms before the existing ones could be eliminated. Our study demonstrated how these mechanisms functioned and how they served to protect certain advantages that an entire class of civil servants enjoyed. In particular, the study indicated that there were no longer good reasons for the NDB to continue its cooperation with the central water ministry as the project-executing agency.

From a pragmatic perspective, the tangible problems of the projects in Baridi and Jamala lay simply in the fact that the plant managements did not have control over the revenues from the waterworks and that the plants suffered as a consequence. In the course of the sector study, however, it turned out that the Ruritanian legal system did allow for the waterworks to retain their revenues. According to an old law of 1965, the so-called Revolving Fund Act, the finance ministry was authorized to release administrative units from the financing system of the public sector. A revolving fund could be established for units that, according to this law, become semiautonomous in financial terms. Revenues would then flow directly into this fund, which the units could administer on their own.

Our strategy of "policy-based lending" allowed cooperation with the Ruritanian government in the water sector provided that it grant semiautonomy as outlined in the Revolving Fund Act. This occurred on July 1, 1994. The waterworks were allowed to place their revenues in a revolving fund and were authorized to use them as they saw fit—initially for a three-year trial period. After this had occurred, we agreed to participate in the water sector again.

Given our experiences with Baridi and Jamala and in accordance with the recommendations of the sector study and the new shift in direction within development cooperation, we also addressed the other end of the process. We insisted that no investments be made in hardware initially, but that first an organization improvement program be financed, which would enable the waterworks to become financially autonomous in a rational manner and to operate and maintain their plants appropriately. In doing so, we inverted the traditional approach and said: For an operation to be able to use an investment positively, it must first demonstrate that it possesses the appropriate structures and instruments.

A second change of course that we were able to push through proved to be equally important. We succeeded in shifting the project-executing agencies from the water ministry to the three waterworks themselves. This represented an effective step in the direction of structural adjustment. It allowed us to circumvent the bloated bureaucracy in favor of decentralization, thereby helping to shift decision-making authority to those institutions that actually had the competence to make such decisions. This also established the prerequisites for the population of the city to develop a sense of self-reliance regarding their own water supply plants through participation. In this way, we emphasized the logic of the market economy and contributed to making the financial cycle smaller and more transparent, thereby promoting a commercial view of water as a public good.

Another dimension of structural adjustment—and the basis of European development cooperation—is the idea that our intervention is just a way of helping people help themselves. Development aid makes sense only as a complementary extension of efforts made by the government, the institutions, and the people of the partner countries. This idea has now become so widely recognized that the Global Coalition for Africa (a high-level political discussion forum founded in 1990) and even the New Agenda for the Development of Africa in the 1990s (established by the United Nations general assembly in 1991) have declared that the issue of *self-reliance*—along with good governance and the coordination of donor support—is central to their efforts. Development cooperation can only exert a sustained effect if it is based on the active participation of the people involved. For these reasons, participation of target groups in the selection, planning, execution, and monitoring of the success of all measures is an overarching principle of Normesian development cooperation.

In the case of the basic and advanced training programs in the plants, active participation means first and foremost that plant management is willing to assume responsibility as the project-executing agency and thus responsibility for the project measure itself. To ensure that the self-reliance is firmly anchored, we have changed the contractual form of the training programs. Up to 1995, we operated in projects of this program according to a so-called *direct procedure*: As the financier and first party in the game, we contracted a consultant as a second party, whom we commissioned to implement a particular development measure for training or organization with a third party, the project-executing agency. For the project in the Ruritanian waterworks in Baridi, Mlimani, and Jamala, we chose a different contractual form for the first time in early 1996, allocating the project according to a so-called *indirect procedure*: As financier, we provided the executing agency with the necessary means to contract a consultant itself. As financier we of course reserved certain rights of control, if only to be able to fulfill our duty of accountability to the MDC. A business management contract with the project-executing agency ensured that we would retain significant influence on the selection and conduct of the consultant. In principle, however, we have incorporated the idea of self-reliance as prerequisite for sustained development into the new contractual form of indirect allocation.

For this reason, the initial phase of the project between the end of February and the beginning of April 1996 proceeded somewhat differently than previously within the framework of direct allocation. The Ruritanian project-executing agencies, represented by the three directors of the waterworks, were now motivated as the principal and employer of the consultant to actively participate in designing the project. Insofar as the three directors played this role from the beginning with great determination and tactical intelligence, we felt confirmed that indirect allocation is the more appropriate contractual form for organizing development projects. Changing the contract also provided us with the added advantage that we are now able to operate according to the same model in all of our projects.

The logic of our work is basically very simple. Once Ruritanians agree to live in cities and to regard certain standards of hygiene and public health as necessary for human dignity, they have to maintain a central water supply and sewage systems to this end. As soon as a society arrives at this point, it is obvious that the wheel does not need to be reinvented. Instead, it makes more sense to draw the technologies and expertise from places where they already exist. If such complex and large technological systems require inexpensive loans or financial contributions from development cooperation, then joint solutions with financial backers must be found. Neither capital nor technologies can be successfully transferred if detached from their institutional framework—we are more aware of this today than when development cooperation first started in the 1960s. In addition, the need to work out joint solutions is based on the fact that the financier is accountable back home to the representatives of the taxpayers, who ultimately pay for this work. The NDB must not only guarantee that this money ends up where it is supposed to, but that it has been used in a rational and effective way.

Implementing this simple logic can lead to astonishing surprises. In regard to our water supply project in Ruritania, however, we can now say that by and large we will achieve our general goals: We have provided an initial thrust in the direction of privatization, deregulation, and decentralization in the Ruritanian water sector. And we have provided initial aid that will enable the three waterworks to deal responsibly with investments.

End of Report by Johannes von Moltke

Afterword : Edward B. Drotlevski

After my interview with Johannes von Moltke at the NDB, I paid a visit to the Ministry for Development Cooperation (MDC) on July 21, 1997. On the first day I spoke with two ministerial officials responsible for conceptual issues of financial cooperation and on the second day with an official in charge of the Africa department. I posed a simple question to each of them: The ministry provides the NDB, as an implementation organization, with tax revenue. This money is earmarked for development policy objectives defined by the MDC. The MDC must of course assume responsibility for the implementation of these goals vis-à-vis parliament and the public. How does the MDC ensure that its policies are actually followed by the NDB? The simple answer that I received was the following: It is not the MDC's job to supervise the NDB. They resolve problems together.

My interlocutors made me feel as if my question were misguided. I was told more or less explicitly that I was one of "those old-fashioned people" who still believed that bureaucracies could be approached in terms laid out by Max Weber. Today, I was told, no one regards bureaucratic organizations as rational systems oriented around the pursuit of a specific objective, but rather as open systems. (An open system consists of a myriad of shifting interest groups that continually renegotiate their aims and are thus strongly influenced by so-called environmental factors.[2]) It seemed like a role reversal for me to hear this interpretation from the mouth of a ministerial official. It is perhaps possible from a scholarly perspective to speak of an open system here. From a legal and policy standpoint, however, it certainly must remain the case that formally organized practices can be controlled, assessed, and predicted.

Other passages from my interview, however, provided deeper insight into the issue of control and assessment. One official explained in detail that the personnel and finances of the MDC were hardly sufficient to execute the ministry's responsibilities properly. At the moment, he continued, there is talk about further cuts in funding. For this reason alone, it is unrealistic to believe that the MDC could examine every project in detail. Another passage of the interview addressed the issue of the NDB's distinctive "corporate identity." In this context, a ministerial department head voiced concern that his coworkers were not always a match for their colleagues at the NDB and as a result allowed themselves at times to be talked into making bad decisions. As he related this to me, he was unable to disguise his respect for the assertiveness of the NDB staff. In closing he even proclaimed, "I'm really a great fan of the NDB."

In yet another passage on the substance of development policy, one of my interlocutors explained what he considered the central dilemma. We hear from all sides, he explained, that wealthy countries such as Normland should allocate a greater portion of their gross national product to development cooperation. The truth, however, is that recipient countries—especially in Africa—are already overwhelmed by the current input. Although the logic of the situation is in fact rather simple, he said, well-meaning people prefer to overlook it. For an input to lead to sustained development and not sustained dependency, the recipients also have to contribute to the process. From the perspective of financial policy, for example, any national economy can only process a certain volume of loans in a positive manner. When this upper limit has been exceeded, the results are negative. On the level of actual individual projects, we find the same pattern. If, for example, the World Bank finances the entire educational or public health sector in Uganda for two years, bringing it up to international standards, and then simply turns the reigns over to the Ugandan government, a new problem emerges. Uganda will not be able to finance this expanded system on its own, and the result will be an even greater dependency than previously. It is unfortunately the case that African governments continually commit themselves to make contributions to their own development, which they are subsequently unable to honor.

Although my interlocutors initially disputed that a problem of oversight existed, it gradually became apparent that this was precisely the issue. Principal–agent theory provides an obvious interpretive schema for this constellation. According to this theory, the relationship between a principal—in the case, the MDC—and its agent—the NDB—is always overshadowed by the fact that the distribution of relevant information is asymmetrical.[3] For an agent to play its role well and to make things easier for the principal in the sense of the contract, it must know considerably more than the principal—otherwise the principal would have no need for an agent. The agent, however, can also use this discrepancy in the distribution of knowledge (which was initially desired and cannot be easily overcome) to its advantage with relatively little danger to its own position. There are two strategies to counter the risk that an agent might not pursue the interests of its principal optimally, and these are best used in tandem. The principal introduces an adequately extensive system of oversight and attempts at the same time to cultivate mutual trust. The balance between these two strategies is in part a question of transaction costs. My impression at the ministry was that in the concrete case of the principal–agent configuration between the MDC and the NDB more emphasis is placed on establishing trust than on exercising control.

2   The Risks of the Entrepreneur

Preliminary Remarks : Edward B. Drotlevski

The Shilling & Partner consulting firm of Mercatoria was commissioned with carrying out the Organizational Improvement Program (OIP) in the three Ruritanian waterworks. The numerous spacious rooms of the converted piano nobile, the many desks, and the professional air of the office manager gave me the impression that I had landed in a large company where business was thriving, money was rolling in, and the big, wide world converged via fax and email. It was only later that I realized not a single desk had been occupied, since other than the managing director, the office manager, and an assistant, there was only one other permanent employee, who happened to be out in the field for a project in Ruritania. In July 1997 the company had no more than ten temporary employees working on short-term projects. The standard interview with the managing director and project leader Julius C. Shilling turned into a lengthy conversation. In August 1997, we spent an additional weekend together on his boat anchored on a lake north of Mercatoria. I will let Shilling speak for himself.

NARRATOR : JULIUS C. SHILLING
Location : Mercatoria, Normland
Date : late July–early August 1997

Acquisition

Sometime in late summer 1994, I was asked by the Normesian Development Bank (NDB) if I would conduct a study on the necessity, options, and costs of an organizational improvement measure for the waterworks of the

Ruritanian regional capitals of Baridi, Mlimani, and Jamala. The NDB has known me for many years as a specialist for those types of questions and I often receive inquiries of this kind.

The NDB is the largest contractor on the Normesian market for all activities in development cooperation. This mainly involves project implementation requiring technical and business expertise. A lot of funds are also spent on expert reports and studies. Last year, 1996, the NDB budgeted roughly 2 billion dollars to finance projects and programs within the scope of financial cooperation. The second largest Normesian contractor is the Agency for Overseas Development (AOD), which received about 1 billion dollars in project funding from the government ministry in 1996. Over the course of decades, an independent market has developed around these two organizations, in which smaller and larger consulting companies vie for projects, whereby the AOD also carries out some projects on its own. One characteristic of this market is that many of the transactions function through networks. There are certainly manifold reasons to set up a network structure, but I find one to be particularly crucial, namely, the necessity for trust and "good will," as American businesspeople call it.

The development business is based on a triangular constellation of financier, project-executing agency, and consultant. This configuration brings to bear a specific logic: Whenever the person paying the bill is not the person benefiting from the product or service, a certain danger of opportunism arises. This danger lies in the fact that the service provider (in this case, the consultant) and the service recipient (in this case, the executing agency) ally themselves against the financier, who cannot easily recognize and prevent such alliances. A typical example: The executing agency is talked into advocating a new water extraction facility that would not be necessary at all if the available water were really distributed instead of seeping out of the pipe system. This increases the consultant's profit and the executing agency receives more fringe benefits. The financier is better able to effectively avoid such precarious situations by establishing long-term mutual trust than by exerting pure control. Considerable transaction costs can be saved in this way.

From the perspective of the contractor, it is also beneficial in this business to work from a basis of greater mutual trust than might be common or necessary on the European market. This is due primarily to the fact that as a rule consultant contracts are concluded beyond the scope of European jurisdiction and the legal situation in such cases is unreliable. Furthermore, experience has shown that non-European contractual partners have different legal conceptions and interpretations of what is at stake and which rules should apply. In our transnational field there is an immense discrepancy between contractually explicated norms and implied norms. More than otherwise, we are forced to rely here on fleshing out the contractual skeleton through noncontractual means, thereby assuring that the other side keeps to the agreed stipulations. To this extent it is also in the interest of consulting firms to establish long-term mutual trust with the NDB. Only with the backing of a loyal financier does it seem at all possible to do business in this field.

Power, however, also plays a significant role in this context. Although the dependencies between financier and consultant are reciprocal, a considerable imbalance develops because the consulting firm is only one among many, whereas the authorities that distribute public funds are an obligatory passage point for all competing businesses. No one who wants to participate in bidding can circumvent the NDB. It is this organization that doles out the money and thus has more influence on the rules than all of the other actors.

Feasibility Study

Diagnosis of the Problem
Every project begins with a so-called feasibility study. For the Ruritanian water project I conducted this study in the fall of 1994 together with an anthropologist and an engineer. The initial problem with the waterworks in Baridi and Jamala had already been identified: The increase in water production that had been brought about by substantial investments in Baridi (starting in 1982) and Jamala (starting in 1973) did not result in any acceptable increase in the volume of drinking water that ultimately reached consumers. In addition, increasing production capacity did not result in any corresponding increase in the volume of water sold. That in turn meant that the waterworks were not financially viable in 1994. Consequently, the first priority was to improve water management. We established five areas of emphasis.

Organization
Our efforts first centered of course on the organizational structure according to which the drinking water was produced, treated, and distributed. A sector study conducted by the NDB had already identified the main problems in this area. The central conclusion of this study was that the waterworks required greater autonomy to be able to operate in an economically viable manner. Even before the project was launched, a preliminary step was taken as a result of the sector study and pressure was exerted by the NDB: On

July 1, 1994, so-called revolving funds were established through recourse to a 1965 law—the Revolving Fund Act—initially for a three-year trial period. In simplified terms, these funds are business accounts that the waterworks received in order to administer their own revenues autonomously rather than transferring them to the finance officer of the regional government within the framework of a public accounting system.

Production, Distribution, and Sales
If there is a water shortage, then more water needs to be produced. We encountered this production ideology on all sides. It required enormous effort on our part to stimulate even the slightest understanding for our interest in certain aspects of business management. Occasionally I had the impression that the three leading engineers with whom we worked most closely would bite the bullet and embrace the project, because they had understood that otherwise the NDB would not offer them any further hardware projects. Our job during the feasibility study in the fall of 1994 was therefore not simply to work with the project managers to develop locally feasible solutions to existing problems. Rather, the first hurdle was to come to a mutual understanding of what the existing problems were.

The management believed that the performance of the Baridi waterworks was dependent on the output figures. Within the scope of the first major technological project, eight new groundwater pumps were drilled and the production capacity had more than doubled since starting up in 1990, to roughly 20,000 cubic meters per day (m^3/d). Since that time the water production of the Baridi works has been around 30,000 m^3/d during the critical phase, that is, at the end of the annual dry period. Based on this figure, at least 170 liters per capita per day ($l/c/d$) are being used in Baridi. The WHO classifies 100 $l/c/d$ as the optimal supply for people having access to piped water free from harmful contaminants, and 20–50 $l/c/d$ as the basic need of a human being. In Normland the average consumption is 145 $l/c/d$, which would mean Baridi lies well above the Normesian level. From the perspective of customers, however, Baridi has continued to employ sporadic water rationing ever since the new facilities were put into operation in 1990, and some urban areas continue to receive no water at all from the works. There is only one explanation for the inconsistency between the two observations: Only a fraction of the drinking water that is pumped through the system ever makes it to the consumer.

In order to assess what is ultimately the most crucial issue—namely, the economic viability of the waterworks—we had to determine the percentage of water that is actually billed. Using the existing data, we calculated that the figure was 60–70 percent in Baridi, 30 percent in Mlimani, and 50 percent in Jamala. Water that was not billed disappeared at two levels. One part literally flowed into the ground due to the numerous leaks both large and small; and another part was not properly recorded by the authorities. While some taps did not even appear in the records, others could no longer be located because the documentation was incorrect. Only about half of the bills issued were actually paid.

The most important point of departure for the project was ascertaining that in Baridi and Jamala approximately 30 percent of the water produced does in fact bring in revenue, whereas in Mlimani the figure must have been so much lower that any statistical calculation would have appeared cynical. I think the figures speak for themselves and were able to convince our Ruritanian negotiating partners that what they had was not a production problem, but a distribution and sales problem. The goal had to be to improve the following key indicators: distribution efficiency, billing efficiency, and collection efficiency. Distribution efficiency measures the share of water that actually reaches the customer relative to the total water produced. (It is therefore the ratio of the water produced minus the volume of water physically lost in the pipe system to the total volume of water produced. As a rule, a distribution efficiency of 75 to 100 is considered acceptable.) Billing efficiency denotes the share of distributed water for which a bill is issued. Collection efficiency indicates the share of issued invoices (or the total amount of issued invoices) for which payment is in fact received.

I presume that the present lack of attention paid to distribution and sales is a cultural relic of the socialist era. Drinking water was nearly free in the Ruritanian cities, and the issue of affordability of drinking water did not come under public scrutiny until 1991. Since that time it has also been said that the neglect of the economic side of municipal utilities was a socialist error. However, in the fall of 1994 the theoretical insight that the drinking water supply is indeed a commercial matter did not yet have any practical repercussions. I had the impression that even the employees of the waterworks at all levels in the hierarchy felt that the idea of earning a living or even making a profit by selling water was rather unrealistic if not downright reprehensible. The effect of this attitude was reflected in a collection efficiency rate of 30 percent and lower.

Realities and How They Are Depicted

The need to present data and analyses on the collection efficiency of the organizations in order to derive appropriate solutions led us to a more fundamental question. Every organization distinguishes between two levels of reality. The first level involves the concrete practices, the materials, and the technologies of the organization; the second level involves the depictions of these practices and things on paper or on a computer screen. One of the most important prerequisites for good management is that this second level be reliable.

This problem is most obvious in the case of the network maps for the three waterworks. The purpose of these maps is to accurately depict the territory—here especially the system of water pipes—so that someone sitting at a desk is able to see where something needs to be done on the ground. The maps of the three waterworks, however, are incomplete and filled with errors, rendering them unusable. The story here seems to be cyclical: With the help of a foreign consultant, the situation is rectified so the maps do in fact correspond to reality. The maps are then used for a period of time, but since reality is constantly changing and the maps are not consistently updated, they inevitably become worthless again at some point. Then a new consultant arrives, corrects the maps, and the game starts anew. Part of this problem is the fact that most people at the waterworks cannot really even work with maps. If cash flow and work times are also taken into account, the situation becomes even more opaque. This makes it next to impossible under these circumstances to make a reasonable decision from an office desk based on written and graphic representation.

Money and Benefits
The problems enumerated thus far are connected to yet another problem. It has not proved worthwhile for civil servants in Ruritania to dedicate themselves to their work in any meaningful way. Over the course of decades they became accustomed to having a secure job, while treating it at the same time as if it were a side job, which did not really provide enough income for them to support themselves. Every family strove to get at least one member into the civil service in order to provide at least a little bit of security for the less secure and more lucrative economic activities of other family members. In the waterworks, whose employees are part of the civil service, any causal connection between what someone does in practice and what that person gets paid for doing it or what position that person holds was eliminated. Status, income, and bonuses depended solely on whether the necessary formalities were satisfied on paper. Whether the tasks were attended to correctly and efficiently, and whether the department ran smoothly—that was a completely different story. Even if things today do not appear as extreme, old attitudes cannot be abandoned overnight.

There is yet another dimension to the elimination of any connection between status, performance, and financial compensation. Whenever a problem arises, people immediately insist that a solution cannot be found due to a lack of funds. Within the old context, this referred to something very specific: The authority that approved the waterworks' budget (a department in the Ministry of Finance) or the one that administered the budget (a department in the regional government) had once again refused to approve an item that had been submitted. Over time this interpretive schema—"lack of funds"—became so fixed in their heads that it did not automatically disappear along with the old budget system.

The best example of this can be seen in the debate about water rates. On the one hand, it is certainly true that the rates that applied in the fall of 1994, even with 100 percent collection efficiency, were too low to cover the cost of operations and maintenance, not to mention the reserves for replacement and expansion investments. On the other hand, however, it is immediately obvious that, given the actual collection efficiency of only roughly 30 percent, the waterworks managers' demand for higher rates was a bit ridiculous.

The popular excuse of a "lack of funds" refers to a problem that affects the waterworks in Baridi, Mlimani, and Jamala "from above." To some extent complementary to this, it is also frequently said that customers do not pay their water bills regularly, which causes the same problem "from below." We were continually told that the unwillingness of people to pay their water bills was a problem of "mentality," that socialism had ruined them by proclaiming that free water for all was a viable and morally desirable goal. It was claimed that the old mentality persisted most stubbornly among the poorer people at the city's periphery, who fetched their water with pails at public faucets, but were unwilling to pay for it. We ultimately discovered that all three cities had large black markets for drinking water. The prices on these markets were many times higher than the official water rates, and a considerable share of the water supply came from the piping system of the waterworks. This can only mean that if there had been continuous running water at the public faucets—which there was not—and if a functioning arrangement for practical and fair payment had been available—which was also not the case—all customers would of course have used the public faucets. We also discovered that the most obstinate nonpayers, who also owed the greatest amounts, were in fact public institutions, that is, the regional and city governments, schools, hospitals, the military, police, and prisons. These two observations suggest how the "mentality" argument actually functioned: The managers of the waterworks hide their failure to collect their water bills by ascribing to the "people" a certain mentality, in this case "poor payment morale."

Working with Foreign Money
Shifting attention to a lack of funds from above and to a lack of payments from below as two factors that management has virtually no influence on is a variant of so-called *black boxing*. People act as if what occurs within the organization has absolutely no consequences and can thus be treated as a black box. The focus of our project, however, is precisely these inner workings of the waterworks. Our Ruritanian contacts initially found it simply inconceivable that we were concerned with anything other than an additional hardware investment. When they finally understood that a redirection of project funds into hardware, such as system piping and leak repairs, was impossible because the NDB, as the financial backer, obviously supported our approach, they developed an interpretation that rendered our approach and their expectations compatible. They designated our approach in general as "training." We should bear in mind here that socialist Ruritania was an intensively "tutored" society: Learning and instruction was constantly addressed, everywhere and always. Merely completing a further training course gave civil servants a better chance for promotion. Thus we were regarded as the new teachers who would set up a classroom somewhere in order to demonstrate to people at the blackboard how accounting, process control, human resource management, and the like functioned. Employees believed that the certificates acquired in the training project would enable them to earn more money. Management imagined that the training units were a kind of "fuel" that we would pump into the black box in order to improve operating results. In addition to the training units, we were also supposed to present management tools that they could implement with the help of this newly acquired knowledge. One thing was certain: The Ruritanians were convinced that we could implement the project only outside the black box, never inside it.

Nothing would be easier for me, as a development contractor, than to unload a set of inputs at the entrance of a black box. It would be a rather safe transaction on my part if I had no responsibility for what came out at the other end. However, the NDB expects measurable results and my success and viability as a consultant is dependent on my ability to achieve these goals. To this extent I am in a vulnerable position. The person who in this case is supposed to have better results at the end of the process is primarily concerned with making sure I don't get into his black box. If he succeeds in this, there is no way that I can help him improve his own results. It is, however, my job to help improve these results, and if this doesn't happen I won't earn any money with the project. For this reason I try to get involved despite his resistance. It is a mixed-up game that occurs only because the project-executing agencies are working with foreign money.

Solution Strategy

The main objective of a feasibility study is to find out if a reasonable solution—in terms of both time and money—can be found for the problems that have been uncovered. We recommended that the NDB finance an Organizational Improvement Program (OIP) that primarily promotes increased collection efficiency, that is, aims at increasing the share of water actually paid for, relative to the share of water billed. To justify this recommendation, we created an explanatory model, in which we identified five weak spots and their causes. The logic of this model can best be illustrated in a graphic representation (see figure 2.1).

The proposed measure focuses on improving the representation of operational sequences, material flows, and other realities on paper or on the computer screen. Thus the main point is the so-called Management Information System (MIS, at the center of figure 2.1). To this end we arranged the introduction of a computer system that was intended to help solve the accounting and documentation debacle.

An information system must in the first place be embedded in a sound organization. The conversion from departments in the local bureaucracy into independent, commercially operating units requires completely new organizational structures. However, the functioning of such units as well as the MIS also requires increased motivation on the part of the waterworks staff, which in turn requires an incentive pay plan and a new personnel management system. In order to be certain that the people do not regard this new system as something artificially imposed, which would lead them to abandon it quickly, we had to avoid readymade solutions. Participation is absolutely essential for any organizational improvement, particularly in development cooperation. In order to increase plant revenue, the payment morale of waterworks' customers ultimately has to be increased as well. One prerequisite for this was that drinking water flowed regularly, which made it necessary to redefine the zones of the pipe system, to install large water meters at the system branches, and to repair leaks. Consequently, maintenance and repairs were important pillars of the project's approach. If these five interventions are intelligently meshed in terms of time and logistics, collection efficiency and thus profitability will increase, in turn leaving more funds for the further

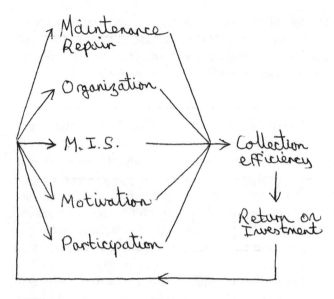

Figure 2.1
The logic of intervention (from Shilling's field notebook).

development of instruments, which will lead to a kind of self-reinforcing improvement of operations.

In order to enable the development bank to better assess its risks, we have listed five critical factors that can be influenced only by the Ruritanian side: (1) The water ministry has to provide the waterworks with additional competent engineers and managers. (2) The waterworks must be permitted to introduce an incentive pay plan that pays employees commensurate to their actual performance. (3) The waterworks must have a right to buy where it is less expensive and to outsource tasks to other companies if doing so makes economic sense. (4) The waterworks must be permitted to develop their own corporate identity in order to shed the image of a state bureaucracy. (5) A binding time schedule must be developed for the attainment of these requirements.

The original purpose of the feasibility study was to identify the best way to convert the three waterworks from government departments into autonomous bodies. This undertaking had to be carried out within a certain time frame. In 1992, the NDB had told the Ruritanian authorities that it would remain involved in the water sector only if the system of financing the municipal waterworks were reformed. This condition was satisfied by the Ruritanians as of July 1, 1994. The waterworks were given permission to conduct a three-year trial run until June 1997 under the Revolving Fund Act. When our feasibility study was submitted in December 1994, it was therefore six months late, so we included a note to the NDB that the program should begin as soon as possible.

Bid

Before the project could actually begin, a number of obstacles had to be overcome. First there was a delay connected with the Normesian distinction between technical cooperation (TC) and financial cooperation (FC). In formal terms, the Agency for Overseas Development (AOD) was responsible here, since the project fell into its main area of business: increasing the performance of individuals and organizations. In terms of substance, however, it would have made little sense to transfer the project to the AOD, since the NDB has been familiar with the waterworks of Baridi and Jamala for about twenty years. Nevertheless, it took almost a year before the NDB received the green light to go ahead.

In November 1995, the NDB finally told us to submit a bid for the project, which we did on December 4. On December 27, we were asked to modify our bid in order to make certain points more precise and then to resubmit it by January 15, 1996, so that the so-called inception phase could begin by late February. The most significant points of the NDB's criticism in late December 1995 were that we had not defined adequately the type of autonomy we were seeking for the Ruritanian waterworks and that our input had not been presented in sufficient detail.

Understandably, all financiers would like to know in advance how their money will be used, in order to help avoid making bad investments. When extending a commercial loan, a commercial bank pays great attention to the creditworthiness of the borrower. Once the loan has been approved, there is not much more for the bank to monitor, since the amortization of the loan as detailed in the loan agreement is the most cost-effective and unambiguous evidence that the money has been invested soundly. Of particular interest to us here is the case in which a borrower hires someone to carry out a project and pays for these services with the loan. The relationship between the borrower as the project-executing agency and its consulting firm as the project implementer can be understood as a principal–agent relationship. The project implementer is the contractor and thus is accountable only to its client, whereas it has nothing to do directly with the financier. Under market-economy conditions, this triangular relationship of financier–borrower–contractor operates according to an official script, which we can refer to as the "O script":

O script:

Financier
⇕
Borrower / Project-Executing Agency (Principal)
⇕
Contractor / Project Implementer (Agent)

The case is very different, however, if there is no market relationship between the financier and the borrower, that is, if the amortization of the loan at a standard market interest rate is not the main mechanism determining the economic rationality of the loan. Under these circumstances the financier is forced to create substitute criteria for determining this rationality. In other words, the financier has to introduce diverse techniques of evaluation, regulation, and control in order to rule out the possibility of the funds being used in an uneconomical manner. In doing so, the financier intervenes in the relationship between the executing agency and the project implementer, which ultimately leads the financier to assume the role of the principal. The borrower, or recipient of the NDB funds, is in principle not bankable, and this compels the NDB to assume the role of the principal, who transfers to an agent—me, the consultant, for instance—the task of delivering a stipulated service to the executing agency. The (unofficial) script of the three-way relationship, which we can refer to as the "U script," now looks different:

U script:

Financier (Principal)
⇕
Contractor / Project Implementer (Agent)
⇕
Borrower / Project-Executing Agency

The first reason that there is a particular regulation and monitoring problem connected to financial cooperation is that we are dealing with aid here and not with business. In the case of our Ruritanian water project, the NDB even went so far as to adjust the legal terms of the contract to be consistent with politically correct O script. For the first time, the bank refrained from directly contracting out a project of this kind. Thus our client—or principal—is no longer the NDB itself, as in all our previous projects, but the Ruritanian waterworks as project-executing agencies. Nevertheless, in the area of monitoring performance, things somehow remained the same and all of the parties followed the U script. Of course, publicly the project had to be presented as if it were proceeding according to the O script.

And as if this configuration were not artful enough, there is yet another catch: the principle of "policy-based lending." Policy-based lending means that someone wants to push through something that the borrowers or subsidy recipients would not do of their own accord. Our Ruritanian water project is a prime example of policy-based lending. The entire project does not seem to make any sense until a development policy stipulation has been included: "If you do not uncouple the waterworks from the state bureaucracy, we will withdraw from your water sector." Since the project-executing agency and its consultant are now working together to do something that the executing agents themselves would not want, it is up to the consultant to compel them. This, however, is not possible because the project-executing agency is officially the employer of the consultants, whom it monitors and pays. In addition, these two parties—the client as a good borrower and the consultant as a reliable contractor—have to prove themselves to the financier. The most obvious way out of this dilemma is for them to band together opportunistically and pull the wool over the financier's eyes.

It is understandable that the NDB will try to prevent this kind of deception by attempting—in addition to its efforts to set up trust networks—to establish as tight a control system as possible. This is particularly difficult, because, in accordance with the approved O script, it cannot be stated explicitly that the executing agency has been moved to accept things it does not actually want. In addition, it is rather difficult to monitor something that cannot be stated on paper. This is the everyday concern for the NDB staff, and it can escalate into an obsession. Thus, the second reason why financial cooperation has a special regulatory and monitoring problem is because we are dealing with politics here and not with business.

Inception Phase

On the basis of our revised bid of mid-January 1996, which was in fact hardly different from the original version, we received the assignment to start the inception phase, despite the fact that we were not able completely to resolve the two critical questions raised by the NDB: What operating model had been planned? What needed to be done to that end? I was able to travel to Ruritania as early as February 27, 1996, together with two members of the future team. It should be said at this point that every project starts with an inception phase, during which it is reviewed whether the goals and procedures recommended in the draft proposal are still realistic under present circumstances. If not, this is the chance to modify the project. If no agreement can be reached at this time, the consultant is still able to abandon the project unscathed. Until the end of the inception phase the financier also retains the right to cancel the project.

Objectives

The waterworks' three-year trial phase, which the project was supposed to accompany, started on July 1, 1994. In early 1996, more than half of the allotted trial period had already expired. In order to make up time, the NDB omitted two steps in the usual procedure. It should actually have sent our offer, as it existed in mid-January 1996, on to Ruritania to allow the future executing agents to examine it carefully and make any necessary corrections. We would have revised our bid accordingly and could then have signed a contract in Urbania. Not until then should the inception phase have begun, which would have been concluded with an inception report amending and finally implementing the contract. Instead we simply began immediately with the inception phase.

When we arrived in Ruritania it turned out that the three project-executing agencies, represented by the directors of the three waterworks, had a different idea of the order in which the necessary steps ought to be completed. The three engineers referred to themselves as UWEs, urban water engineers. From the perspective of the UWEs, the contract should be signed at the end rather than the beginning of the inception phase. At the time I had no idea that the NDB had given the executing agents four weeks to revise our draft project, which overlapped with the four weeks of the inception phase. In doing so, the financiers had reversed the usual order of contract signing and inception phase without informing me. This deliberate secrecy inevitably led the UWEs to regard my interpretation of the order of contract signing and inception phase as an expression of entrepreneurial audacity. And I, on the other hand, could only view their interpretation as naive chutzpah.

In addition to the legal differences with respect to signing the contract, a substantive discrepancy also emerged. In March 1996, the NDB said it no longer expected that the objective of the project would be to support the waterworks in implementing the Revolving Fund Act, as had been originally agreed upon. Instead, the objective was now to disengage the waterworks from the public utilities entirely. This was not possible within the existing legal framework, which had been amended in 1994 precisely for this project. The NDB, in other words, classified the project as a pilot project that could only function outside of a Ruritanian legal framework. At the same time the NDB expected the executing agencies to be instrumental in developing the project design. In substantive terms, we had no problem agreeing separately to both of these NDB expectations. In terms of strategic negotiations, however, we found ourselves in a stand-off, especially since we still did not have a contract.

From our negotiation partners' perspective, neither of us was entitled to debate Ruritania's government policies, much less make decisions about Ruritanian affairs. The water ministry, which they regarded as the highest authority, had reluctantly allowed them to act as the project-executing agencies. This was something new nationwide and was considered rather scandalous in the capital Baharini. For this unusual autonomy, a fixed framework was staked out, which, from the ministry's point of view, was appropriate to the waterworks' authority and sphere of responsibility. It was their job to decide, together with us, which systems and training measures were most urgently needed in taking optimal advantage of the semiautonomy granted through the revolving fund. Everything else was the responsibility of a ministerial committee. An amendment bill submitted by the minister for a vote could supposedly be reckoned with any day, at the very latest by the parliamentary session in April 1996.

In March 1996, there was no one in Ruritania officially representing the NDB's position that this was a pilot project seeking new, previously unheard-of procedures for organizing municipal waterworks in Ruritania. Nevertheless, it was our task not only to implement this impossibility, but to do it on the basis of a partnership. The more obvious it became to our Ruritanian contacts that the objectives in our heads were completely different from theirs, the less inclined they were to sign the contract. On the other hand, had we distanced ourselves more from the expectations of the NDB, it is almost certain that the Normesian side would never have signed the contract.

Although there is no theoretical solution to a problem like this, it can nevertheless be resolved in practical terms. We agreed with our negotiating partners that the ministerial committee to whom they repeatedly referred was in fact responsible for determining the precise legal status of the waterworks, which would be autonomous in the future. We also stressed that as soon as the amendment to the corresponding law was actually passed in parliament—that is, possibly as early as April of that year—a second step would immediately follow. The now autonomous waterworks would have to be reconstituted as economic agents, and that would involve considerable work, which would certainly not fall within the jurisdiction of the ministerial committee. They seemed to accept this division of the task, and together we developed a constitutional framework for the waterworks. It was necessary to keep the simulation very vague at those points that were dependent on a legal framework that did not yet exist.

Areas of Responsibility

An important prerequisite for the success of the planned Organizational Improvement Program (OIP) was defining the technical assignments. One cannot expect good payment morale from a customer who has access to water only for a couple hours every few days on an irregular basis, and who otherwise turns on the faucet and hears nothing but wishful rattling in the pipes. It was thus obvious that an increase in collection efficiency—that is, the share of billed water that is actually paid for—was more likely if at the same time the physical leaks in the pipe system were repaired so that more water would actually reach the customers, who would then be more inclined to pay their bills. To get this mechanism going we listed a series of technical assignments within the scope of our OIP that could be executed without any major hardware investments.

This seemed necessary at the time we were writing the draft proposal, that is, in the fall of 1995, for the simple reason that the NDB was then pursuing a strategy of not financing any additional hardware for the Ruritanian sector until it had undergone comprehensive reform. It was therefore crucial for the Ruritanian waterworks to improve the technical situation through their own resources in order thereby to become "creditworthy." This also provided us with a favorable starting condition since the UWEs (urban water engineers) were as a result strongly motivated to work together with us to bring the OIP to a successful conclusion. For this reason, one of the largest items in our bid was for a water engineer for roughly 20 person-months, who was to work in this area.

The situation changed fundamentally during the inception phase. This was triggered by the fact that the NDB tacitly abandoned the deal it had made—that is, to first achieve tangible success in the OIP before making additional hardware investments—as if the issue had never been discussed at all. Parallel to the OIP, a technical program (TP) was now also planned. This change in strategy initially caused the UWEs to suddenly lose their greatest incentive for successfully implementing the OIP, since the change confirmed precisely what the Ruritanian engineers had believed all along: that technology is the most important element; only when the technology was adequate could one even begin to concentrate on secondary aspects such as process control, cost accounting, billing, motivation, image, and the like.

The simultaneous planning of an OIP and a TP by different players of course also gave rise to a number of problems regarding areas of responsibility and jurisdiction. Reorganizing the zones and installing large water meters at the major distributor points in the system was temporarily interrupted by the UWEs in light of the pending TP. However, the waterworks' service during the OIP period could not be sufficiently improved without these measures. Consequently, the prime objective—to increase collection efficiency—could no longer be achieved. Eliminating the leaks was also postponed because it was believed that the TP would provide more generous solutions. The OIP, however, continued to be responsible for training the crews searching for leaks. In addition to the negative effect on the service, I also envisioned a satire on the horizon: Teams with high-tech sonographic devices would attract onlookers on the streets of Baridi, Mlimani, and Jamala, and these fascinated observers would be amazed to discover that, after successfully locating a leak, the teams would contentedly move on in search of the next one.

One problem regarding the areas of responsibility that has plagued us since March 1996 was related to customer data. At the beginning of the inception phase it appeared as if this responsibility was to be shared among all the parties involved, without any clear understanding about who was supposed to do what or precisely what was even required. The executing agents, at least the UWEs from Baridi and Jamala, maintained that all the customer data had already been collected. From their perspective the important thing was to get all the already existing data transferred from paper to the computer and have it integrated with the old data. During the inception phase we did our own review of the data from Jamala and Baridi and came to the conclusion that the data were incomplete, filled with errors, and inconsistent. Because the customer master file would be the database for the software package that we were to develop, we suggested that we assume the responsibility for the project data. But the UWEs rejected this solution. As a result, we remained in charge of the software package, and the customer data of the three waterworks remained their responsibility. This marked the beginnings of another satire: The project could ultimately lead to more and better invoices being printed without increasing the ratio of invoices paid.

Facts and Figures

Our third task during the inception phase was to break the entire project down into steps that could be operationalized and calculated. This was not easy, because—corresponding to the nature of the business—two elementary prerequisites had not been satisfied: Neither the objective nor the areas of responsibility had been established. On top of that came a third problem due to the nature of the development business.

The main weakness of the organizations that are supposed to be developed (such as the waterworks) is generally the fact that the data on the work processes, material flows, and cash movements are largely unusable. At the same time the consulting firm hired to deal with eliminating this problem has to provide the financier with facts and figures in order to justify the ensuing costs. Most of these facts and figures can be provided only by the executing agency, but are either missing or unreliable, which is why the consultant was sent for in the first place.

As was the case with the previous obstacles, here too there was no appropriate theoretical solution. In practice, however, a simulation is usually prepared on the basis of the existing data. Attempts are made, of course, to review and correct this data, but that is possible only to a certain extent within the scope of feasibility studies and inception phases. For example: If there is no reliable calculation system that is routinely built in, which can be used to calculate the cumulative debt of all customers, then the only way to achieve this figure would be to individually record each customer account by hand. First of all, the necessary time and expense to record this for several thousand customers would be totally out of proportion to the value of such a procedure; and second, at the latest by the fifth customer account, it would become obvious that the numbers entered are nowhere near reliable. Correcting this data would require a considerable outlay of both time and money. In search of firm ground, one ends up precisely in the mess that the entire process was supposed to eliminate. So there is nothing else to do but make an educated guess. The difficulty is that once you accept a bid based on such data, you are stuck with it. It often turns out that the situation is even worse than the original figures made it seem. Consequently, the project gets more and more expensive over time. Although this rule of thumb is familiar, it is impossible to assume a worst-case scenario and add another 25 percent right from the start because as the consultant you are forced to make a competitive offer.

Because of this data situation we were unable during the inception phase of our project to find out what services our client wanted us to perform.

Instead, we were more involved with defining the overall purpose of the project. In doing so, we were confronted with contradictory expectations, had to wrestle with unclear areas of responsibility, and did not have any clearly ascertainable point of departure to fall back on. Under these circumstances our proposal could hardly be more than a simulation that would be translated into a realistic plan only in the course of the project itself. However, any official acknowledgment of our contribution is a different matter, in which only the O script can be taken into consideration. Seen in this light, the story is very simple: In the course of the month of March, we (that is, the three UWEs and the three people from S&P) went through all unclear aspects of the January 1996 version of our draft proposal. The corrections we worked out together were recorded in the so-called inception report. On April 4, 1996, the UWEs, that is, the executing agents, then signed the two documents simultaneously. This served both to define the project and to conclude the inception phase. We left Ruritania two days later and sent the documents to Urbania.

Inception Report

Our main concern was now whether the bank would accept the detailed changes that we had made together with the UWEs during the inception phase. It was basically a matter of establishing more precisely the legal and operational forms that the waterworks were to assume after the Revolving Fund Act guaranteed them the right to set up an account of their own and a revolving fund. We had developed the aforementioned model simulation in order to clarify this point. We also determined what measures needed to be taken for the organizational improvement. Although we could not determine once and for all what type of business was desired—since the NDB rejected the only possible legal form—and could not resolve every question down to the last detail, we were still able to draw up a relatively detailed catalog of measures.

Catalog of Measures and Action Plan

Together with the UWEs we divided up the Organizational Improvement Program for the three waterworks in Baridi, Mlimani, and Jamala into eighteen separate measures and arranged them in a catalog according to usual procedures (see a sample page of the list in figure 2.2 to get an impression of this kind of project document). This table facilitates understanding the nature of the measures and the logic behind the intervention. Each horizontal row records five aspects of an individual measure. The first column is the identification code; the second column ("description") explains the area of intervention more precisely; the third column describes the initial "problem"; the fourth gives a possible "solution"; and the fifth column ("inputs required") records the necessary outlays for the action. Experts examine the table, looking first at the fifth column to find the individual measure with the highest input. This usually denotes which row in the table is the most important, that is, which problem in column 3 is considered the most urgent. The logic of the draft program then emerges based on this. In our case, we would quickly come to Code 009, the ninth row in the table, and thus to the aforementioned problem of distribution. Toward the end of every project, the main focus is as a rule again on column 5. The so-called negotiations for supplementary work then deal with determining whether or not the planned inputs were sufficient to achieve the agreed-upon objective, if they were really necessary, and if they were implemented optimally.

As regards consultant contracts, the supplementary negotiations usually deal with defining interfaces. If the objective, the definitions of the situation and the problem, the approach to a solution, and all the related data turn out to be inconsistent, then everyone who is expected to pay unscheduled expenses will say: "It wasn't my fault; you're responsible." For this reason the interfaces normally are at issue, and this is why the next table that must be included in every project proposal or inception report, the action plan, is also essential.

The action plan goes a step further into the details, approaching the concrete actions that are to take place in practice (see a sample page of the action plan in figure 2.3). The horizontal rows in this table do not list operational areas, but instead identify actual, individual actions that are defined precisely according to content, scope, and areas of responsibility. The month columns in this table translate the duration of work per activity into calendar time, making it easier to recognize how an activity is carried out over time. This is depicted on the table by means of a bar that extends over the appropriate number of month columns in each case. The bars have different shading depending on the area of responsibility. Viewing all the bars as a whole depicts the planned time period for this project and shows the chronological overlapping and meshing of individual actions. Almost everything that needs to be known about a project is recorded in this table.

More clearly than in the catalog of measures, it is possible here to grasp the significance of defining the interfaces. Two lines were especially inter-

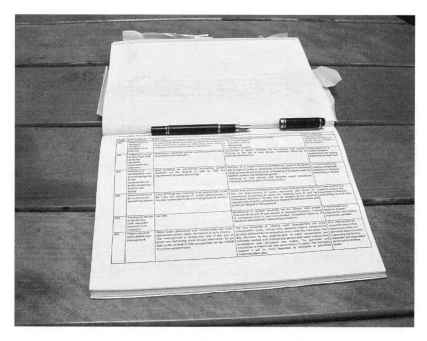

Figure 2.2
Catalog of measures (sample page).

esting as regards our project: (1) customer data and reorganization of the zones, and (2) ongoing system maintenance and stepwise improvement of data reliability. In an Organizational Improvement Program it is absolutely impossible to restructure the entire internal organizational data system. One can never assume a clean slate. Not a single elementary data type that is needed can be corrected without recourse to other existing data types and preexistent operations. It is also impossible to avoid referring back to external databases, such as city maps, urban development plans, network maps of other line operations (telephone, power), urban localization techniques, land register maps, postal address systems, telephone books, and the like. For this reason, the responsibility for the basic data upon which everything else is structured never lies exclusively with the consultants. They always have to have a basis for their data, even if they know that this is possible only to a limited degree. This makes it all the more important to define the areas of responsibility. In the two rows mentioned here we had clarified responsibilities in a delicate area: The executing agency was responsible for the

Figure 2.3
Action plan (sample page).

customer survey and successive improvement of data reliability; we were responsible for zone reorganization and continued system maintenance.

To succeed as a central control instrument of a project, an action plan also needs so-called milestones. These serve to draw attention to the overlapping of individual activities, as depicted by means of the bars on the plan. Some cannot be started until others have been completed. This interdependence is of course particularly critical in cases in which an action lies in someone else's area of responsibility, so that one's own success is dependent on the accomplishment of an activity over which one has no control. Our first milestone for July 1996 was: "Assure that Ruritanian legislation provides for sufficient freedom of action for urban water departments." Put as diplomatically and flexibly as possible, this meant that the autonomy of the waterworks had to be regulated politically and legally by July 1996 in order to be able to restructure aspects of the project that required a legal framework. This included in particular an incentive pay plan. Our second milestone, for June 1997, was: "Remuneration is adequate in relation to economic capability and

Figure 2.4
Statement of accounts (end of fourth quarter, June 30, 1997).

labor market conditions." For us as consultants to achieve the second milestone, the introduction of incentive pay needed to be legalized a year earlier; otherwise the second milestone would be illusory.

In order to record the entire reality of the project, in addition to the catalog of measures and the action plan, a list of indicators is also needed to determine whether or not the objectives have been accomplished. An overarching goal serves as the point of departure. In our case: "The urban population should be supplied as needed with sufficient quantities of clean and healthy drinking water at the lowest possible cost." This political objective could be translated into a technical-economic project goal: "To establish the waterworks of Baridi, Mlimani, and Jamala as autonomous, financially self-supporting units that are capable of surviving economically in the long term." We agreed on ten indicators, three of which depicted an improvement in water distribution. Commercial losses were to remain under 10 percent. Conversely, the billing and collection efficiencies were each to be raised to at least 90 percent.

The three main indicators measured waterworks performance that could only partially be influenced by the project. They were largely dependent on the executing agents and other players. This was especially obvious as regards collection efficiency, which primarily had to do with whether the customers could also be localized and contacted. This in turn was largely dependent on the reliability of customer data, for which the executing agents had requested and assumed responsibility. Increasing collection efficiency was also dependent on the legal system, which at that time did not offer any effective instruments for dealing with illegal customers.

Which Script Applies?

In early April 1996 the NDB received the modifications to the draft proposal that we had drawn up with the UWEs during the inception phase in March and recorded in the inception report. According to the financing and agency contract between the NDB and the executing agencies, the signature of the financier was needed for the contract to be valid, and it was still outstanding. This signature never arrived. Instead, in mid-May we received a statement from the NDB demanding that S&P submit a thorough revision of the report as an additional prerequisite before the contract could be concluded.

First, the letter was directed only to S&P. Thus the consulting firm was addressed as the sole author of the inception report. This is contrary to fact, because we had drafted every single sentence of the report in a protracted process of negotiations with the UWEs. It was the NDB itself that in 1995 had set up the situation so that the three UWEs were in charge of the events, and we had to offer them something, in exchange for which they would either give us the contract or not. So in March 1996 we had had no other option but to write the report together with the UWEs. Now the NDB still did not approve the inception report, but instead wanted it to be more consistent and binding in terms of privatizing the waterworks; they also demanded more binding precision in implementing the aims of the project.

Were the NDB to treat the executing agents as coauthors, but refuse to accept the report, this would be inconsistent with the O script—according to which the project-executing agency is the sovereign principal that is offered a loan. It was the NDB itself that had already agreed with the Ruritanian authorities back in 1992 to finance the OIP as soon as the Revolving Fund Act went into force. Officially, therefore, there was no way to make any further modifications, and the conditions had already been satisfied in July 1994. The only way to reconcile the reliability of the bank and the autonomy of the executing agencies, on the one hand, with the bank's dissatisfaction about the project proposal, on the other, was to reinterpret the authorship of the proposal. That is why the NDB now treated the consultant as the sole author of the inception report.

The NDB's letter dated May 1996 was remarkable for yet another reason. At that time it was known that the technical program (TP) would take place in three phases. In the first phase the pipe systems were supposed to be analyzed and mapped. The data received were to be used in the second phase to draw up the specification of terms. Not until the third and final phase of the TP were the commissioned construction companies supposed to repair and renovate the facilities. This would include, finally, repairing the leaks. This is a logical order from a technical perspective. But it had meanwhile become known that the start of the third phase, that is, the repairs, was still up in the air, because its financing had become doubtful. Nevertheless, the NDB's letter to us stated that the OIP was not "to facilitate the immediate start of leak repairs." That was the official sanctioning of the parody of a high-tech leak detection program that is content to locate leaks in the pipes only to continue the search for more of them. It also stipulated that we could not address one of our important activities—reorganizing the zones in the network and installing major water meters at significant branches in the network—until we had received new network maps from the TP. To avoid the belly landing that seemed inevitable, we were supposed to coordinate efforts with the two engineering companies that carried out the TP.

There was an urgent need for coordination in the area of interface management between the OIP and the TP as a result of the internal departmental logistics of the NDB. Coordination of this interface management was delegated in accordance with the O script to the UWEs, who as executing agents would bear the responsibility. According to the U script, the NDB expected the coordination effort to come from the consultants, because it knew very well that the UWEs would not be able to manage it. But there are three consultants: two independent and competing engineering firms that carry out the TP and our company, which is supposed to be responsible only for organizational improvement. None of the three companies was ever commissioned to do coordination, and none was allotted the means and information that the job would require.

Project Launch

Although the NDB had rejected the inception report with their letter of mid-May 1996, it simultaneously approved of two project staff members from

S&P doing work in the field. Two so-called long-term experts started their work on May 25 in Mlimani, where they set up the project headquarters. As with every project, the initial concern was dealing with practical things. Equipment and materials had to be ordered, workplaces set up, and telephone lines established.

Clarification 1 : This Is about Model Transfer

As these things were running their course, we submitted a second version of the inception report in June 1996. In response we received a statement in early July from the UWEs; it was a replica of the NDB letter of May 1996, down to the smallest details in style and tone. The presumptuousness with which the three directors of the waterworks not only tacitly accepted how NDB shifted the authorship to the consultant, but even openly agreed with it, was glaring. With their criticism of the inception report, the UWEs were acting as if it had never been their own text. This accomplished one thing in particular: Their responsibility no longer lay in developing—together with us—an as yet unknown model that would be adapted to suit their own local conditions. Instead, their responsibility now lay more in their monitoring us as we installed a universal model that could be taken from current handbooks on management. Whenever anything goes wrong, it is practical to know immediately who is responsible, since according to the handbooks there is always only one responsible party and the model to be installed had already been determined. Corresponding to the O script, the UWEs thus assumed the role of the principal who was criticizing the unsatisfactory result of its agent.

This shift in perspective was also in line with locally tried and tested notions in yet a second respect: In Ruritania the story has often been told that many of the country's evils can be traced back to the greed and incompetence of foreign consulting firms. The point is that here—in contrast to Europe and the United States—not the World Bank and the IMF, much less the NDB and other national development banks, become the scapegoats, but instead the consulting firms that implement the projects. This interpretation fits right into the Ruritanian view of the contractor as a rogue who survives by defrauding decent people.

Clarification 2 : The Consultant Knows the Facts and Figures

The critical statement by the urban water engineers in July 1996 also corresponded with the views of the NDB in yet another way. They demanded that we consultants include more precise data about their own plants in the jointly prepared inception report. I have experienced quite a few real-life parodies, but this one outstrips all the others by far. The letter from the UWEs asked us for statistics about the following matters: (1) water rationing; (2) the percentage of consumption measured by water meters; (3) water quality; (4) estimated water loss through leaks; (5) amount of irrecoverable debts through nonpaying customers; (6) estimated billing and collection efficiency; and (7) assessment of the quality and progress in recording customer data "in accordance with your expectations."

Everyone involved has known for years that the lack of precisely these figures is one of the key problems of the waterworks. This problem has been publicly recognized at the latest since the sector study of 1992. Our feasibility study of 1994 confirmed and specified the diagnoses of 1992. At that time neither side had ever disputed that the main problem of the waterworks was that a large share of the water flows to customers to whom contact has been lost simply as a result of the documentation chaos. As recently as March 1996, during our actual work with the UWEs on their desolate lists and incorrect figures, we would joke with them about the fact that almost every figure was erroneous and that it was virtually impossible to review any of them without falling ever deeper into the statistical quagmire.

What was happening here? First, the three UWEs, along with their predecessors, had contributed to the fact that their data systems had more or less completely lost touch with reality. As a result, they reduced their operations to a bare minimum, so the NDB initially ruled out any additional hardware investments. Then the NDB promised them computer systems and an expert—together with all the accompanying measures for 3 million dollars—in order to get their data systems back under control. When the measure was completed, according to the agreed-upon objective, it was supposed to be possible again to produce accurate data. Now the same UWEs were playing the brand-new executing agents, proudly commissioning a globally operating consultant and giving instructions to "quantify! How much for each town?" at the start of the project. That is a hitherto unsurpassed parody of the cooperative effort.

But even this parody had a rationality of its own. Based on the O script it was not advisable to acknowledge the true extent of the unreliability of the data. The UWEs would thereby undermine their reputation in Baharini, where the project reports are read. Since that remained the place where they make their careers, this could not be in their interest. Also with respect to the NDB, they would hurt their credibility as executing agents if the full extent of the data swamp were to become known there. Finally, it was also not in the interest of the NDB to let the hopelessness of the situation come out into the open, since the feasibility of the project would then appear doubtful.

Because I saw through that from the outset, I too was willing to show some restraint. In the inception report we concentrated more on the solutions and the tasks to be accomplished than on the misleading data. For example, we merely hinted at the fact that only about 300 of about 7,000 customers in Mlimani regularly pay their bills. And we refrained from mentioning the main problem—the dismal situation of the customer data—after our attempt to have this issue put into our area of responsibility failed last March due to resistance by the UWEs. In my view, part of the harmony during the inception phase in March developed because we did not include all the details in our report, though we did speak clearly about the miserable data situation. I understood this as a necessary diplomatic deal between the UWEs and myself as the consultant in order to be able to get the project going at all. The NDB's letter in May 1996 gave the UWEs a new way out of the dilemma, which they evidently felt was more promising for them: Blame the consultant for failing to collect the necessary facts and figures.

The three negotiating partners were already so ensnared with each other in August 1996 that no one wanted to think about the contract that had not yet been concluded. In response to their criticism in July, we wrote to the UWEs and made crystal clear that the only facts and figures in the report that we would change were whatever they would send us themselves. Otherwise we would consider the preliminary studies to be finished in order finally to concentrate on the actual project at hand. As could be expected we did not receive any answer at all, much less new facts and figures. By remaining silent about this and thus leaving the UWEs to flounder on their own, the NDB signaled to us, according to the U script, that from their standpoint the project was no longer at issue. There was also no longer any reason whatsoever to formally confirm the start of the contract insofar as the NDB had already declared it valid when they transferred the funds. For the sake of formality we submitted a final version of the inception report in September, which was filed without comment. The official startup of the project was retroactively set at July 1, 1996, so that the first quarterly report covered the period from July to September 1996.

Clarification 3 : This Play Is Being Performed with Two Scripts
According to the task and the plan, we had to start reorganizing operations in the first quarterly period. Since "milestone 1" had not yet been achieved from the Ruritanian side, which meant there was no legal foundation for the

reorganization process, we were slowly falling into a trap. In August 1996 we were commissioned by the NDB to organize a strategic planning workshop on the bank's behalf.

The divergent interests that were supposed to be harmonized were meanwhile relatively familiar: The NDB wanted to turn the OIP into a pilot project that would seek a consistent path toward privatization and would test models for it. The UWEs wanted to use project training to obtain management tools and computer networks and other than that wanted to be left alone. The representatives of Ruritanian politics and the public administration wanted to push through the project as it had originally been agreed upon within the framework of the Revolving Fund Act. And for us, doing our job as consultant hinged on our being informed by the end of the workshop what the legal form of the waterworks would be in the future. In preparation for the workshop we distributed a discussion paper which outlined the concrete problem we had to solve: If the present Revolving Fund Act did not guarantee the waterworks sufficient autonomy, yet this was a prerequisite for the project (milestone 1), how were things to proceed?

The workshop took place in Mlimani from October 31 to November 2, 1996. The evening before the workshop started all Ruritanian participants met on the terrace of the Grand Hotel. My project anthropologist Samuel Martonosi and I happened to walk past their table on our routine trip to the bar and were invited to stay for a beer. The undersecretary of the water ministry was seated at the head of the long table. As the highest-ranking official there, he was the natural spokesperson for the group. With a nod in our direction he let us know that they had already agreed upon one of the models we had drafted. This agreement on a common position was remarkable to the extent that the parties represented definitely had different interests. In preparation for the staging of a Ruritanian–Normesian workshop, all internal differences had apparently been cleared up. Even more remarkable was the unspoken manner in which we were automatically treated in this situation as "the Normesian side." It seemed that the group here had totally forgotten that according to the O script, we were actually working as consultants for the three UWEs and were therefore commissioned by them to do the job they defined.

In the other hotel, from which we had just come, the Normesian side was having their dinner; the preliminary discussion there was less harmonious. For simplicity's sake I have until now always spoken of "the NDB" as a unified actor; however, some differentiation is in fact necessary. The NDB delegation was made up of three members. The two representatives from the Countries and the Technology Divisions were following the O script that evening, emphasizing that the main purpose of the workshop was for the Ruritanian participants to work out a model suitable for them in discussion with the other parties. The financier was to assume the role of a listener in this process, only intervening if it could not accept responsibility for the decisions made. From their perspective it was more about tangible aspects of the project than about the legal framework.

The representative of the Sectoral Division, Mr. O., on the other hand, assumed that it was already rather obvious which model was best. Like the Ruritanian side, he too had forgotten that according to the O script we were actually working for the UWEs. Instead, following the U script, he accepted without question that we were to promote his concept, which differed in two aspects from the one favored by "the Ruritanians": First, Mr. O. envisaged the legal form of a private limited company, which would fall into the legal framework of a company ordinance, cutting off all connections between the waterworks and the civil service. Second, in addition to private participation he also wanted the city to be a coowner. To the extent that Mr. O. was the leader of the delegation and his two colleagues kept a low profile in the discussion, his conception was in fact treated as the NDB conception for the duration of the workshop.

The next morning officially started with a baroque exchange of pleasantries and the diplomatic marking of unalterable positions. Not until the afternoon did discussion gradually get down to the actual matter at hand. The undersecretary declared that in Baharini the decision-making authorities had already agreed on a model called the "Urban Water and Sewerage Authority," which corresponded to the consultant's model of "the profit center." He said that a ministerial task force established specifically for this purpose was working on the details at this very moment. However, when requested to explain some of these details, he declined, noting that he did not want to anticipate the results and that he did not know the details anyway. As he spoke, two unknown members of this task force sat there without identifying themselves. He spoke with the self-evident authority of an undersecretary of a sovereign country and thus implicitly declared the subject of the workshop as we had prepared it on behalf of the NDB for irrelevant.

At the same time he did not want to antagonize the NDB, so he remained friendly and seemingly willing to negotiate. He explicitly welcomed the unique opportunity for a fundamental discussion on reorganizing the water supply. Mr. O, the NDB spokesperson, in turn, also expressed his pleasure regarding this openness for discussion. But after that he made hardly any serious moves to promote the privatization of the three waterworks. With amazing ease, agreement seemed to be in the offing between the two main contracting parties. They were well on their way, he said, and now could calmly discuss a few further interesting details.

The conversation that was able to develop after the statement by the under-secretary and after the NDB's agreement had to remain rather far removed from reality, in order not to threaten the achieved harmony. The moderator realized that very quickly, and didactically tried to bring about a situation that would function according to this pattern: "Now we'll all try to pretend that we are the body responsible for making a decision in this matter. What arguments would we bring together?" A few hours were spent exchanging ideas about the configuration of the board, the separation of political regulation and economic management, and the role of the city and the customers. But the words spoken here were merely wishful thinking.

From a logical standpoint, only two solutions could really be considered. Either: The project would be postponed until the Ruritanian side satisfied milestone 1 by putting a legal and political framework on the table that allowed the concrete aim of the project, as the NDB saw it, to be implemented. None of the gathered parties wanted this solution because everyone had a primary interest in seeing the project continue. Furthermore, an interruption would have caused the Ruritanian side to lose face insofar as it would show that no one believed a word of any of the continued promises regarding milestone 1. Additionally, the group gathered in Mlimani would not have the authorization to make this decision anyway. Ultimately, this solution would have also threatened the NDB with a loss of face in a different area. It would presumably have been quite a disgrace if it had to admit to the AOD (Agency for Overseas Development) and the MDC (Ministry for Development Cooperation) that the project that now had to be interrupted was precisely the one that took a year of struggling before finally receiving special authorization.

Or: The currently valid legal and political framework would be accepted. All parties were aware that the decision-making process in Baharini could drag on for quite a while. At the same time it was clear that the project only had twenty months of running time left and it definitely needed a precise framework for this period (November 1996 to July 1998). This contradiction would make it necessary to separate two things—the project and the public sector reform. However, this solution would have meant a great loss of face for Mr. O, the spokesperson of the NDB delegation, to the extent that his contribution to the compromise had been to steer toward the "water authority" model. Now it would be awkward to negotiate him back down to the status quo that had already existed in 1994.

These contradictions ruled out the only two clear and obvious solutions. Based on the logic of diplomacy and the workshop's power structure, a third solution emerged. Everyone agreed on the concept of a pilot project. New forms of decentralization and deregulation of public services would be tried out, which could later be useful in the national reorganization of the water sector. The water authority model was confirmed, at the same time reiterating the old promise—our milestone 1—that the new legal and political framework would be on the table six months down the line, by April 1997 at the latest.

I could not believe my ears when I heard this promise. The workshop had been held in the first place because this promise (milestone 1) had not been satisfied in time, that is, in July 1996. At the last moment, when the other participants were already packing their bags, we managed to indicate that the project could not continue under these circumstances. The undersecretary then declared his willingness to issue special authorization to bridge the period until the law was amended, since we as consultants would already have to begin taking action. Agreement was even found regarding an already known list of obstacles that would surely appear on the agenda very soon. This list contained the problem areas that we had already mentioned in our feasibility study in 1994: In order for the project to have a chance, an incentive pay plan would have to be implemented, personnel transferred and in some cases laid off, and new, skilled employees hired. And the waterworks would have to be permitted to make purchases nationally and internationally on the free market. The undersecretary explained in the presence of the NDB delegation that these obstacles could be removed. And he added that his office was always open to us, the consultants.

Thus the workshop ended with two improvements over the situation at the start: First, it was publicly known to some extent that the project was set up for maneuvering, and the undersecretary had committed himself on this point—also vis-à-vis the NDB. Second, the critique of the inception report that the NDB expressed in their May letter had been refuted, since it was impossible, even in this group, to be more specific.

In principle, though, we were still supposed to privatize something that could not be privatized. This quandary had simply become a bit more transparent: In contrast to the UWEs and the NDB, we had already believed for quite some time that decoupling the waterworks from the government bureaucracy could hardly be enacted by the water ministry on its own. Consequently, we had also invited representatives of the Ministry of Finance and the Prime Minister's Office to the workshop. Insofar as these guests neither appeared nor declined the invitation, they made one thing crystal clear: The issue of nationwide reorganization of the civil service and the regional government is negotiated elsewhere. From this perspective, it was apparent in regard to the list of points for which the waterworks required special authorization that the undersecretary in the water ministry had promised a variety of things that did not even fall in his area of competence. Personnel and purchasing, for example, are the responsibility of the regional government and subject to general civil service guidelines and are therefore in the area of competence of the Prime Minister's Office and the Ministry of Finance.

Several months after the workshop we did in fact receive a copy of the bill supplement that the Water Ministry had forwarded on January 3, 1997, for legislative review. Since then we have known what some participants must have already known during the workshop in October 1996: Part IV of the bill contains the often-cited amendment to the Waterworks Ordinance, which had been long in coming. The relevant sections here, 32 and 33, say that the minister can establish an autonomous water authority and commission it to supply a certain area with water. Regarding the issue of what a water authority is supposed to be, the definition was cryptic, simply stating that it should be legally regarded as a corporate body that can sue and be sued. A precise juridical, organizational, and financial definition of this point, however, had been the linchpin of the talks since December 1995. But this law could not provide such a definition, because it simply did not fall within its sphere of jurisdiction.

At the Mlimani workshop in October 1996, therefore, a gathering of the highest officials confirmed that the wrong tree had been barked up in 1995, and they had been wasting their time since then, waiting for the amendment to the Waterworks Ordinance to present a solution. One might infer that this pseudo solution had been arrived at either out of craftiness or ignorance. I don't want to believe either one, however. I cannot accept that a Ruritanian undersecretary could be so incompetent or that a representative of the Normesian Development Bank could not know what institutions are responsible for privatizing a department of the civil service. Instead, I believe that in Mlimani the main interest on both sides was to allow the project to continue no matter what the circumstances. This is why they were playing for time. Only the UWEs and we, as consultants, could have suffered any damages from that, but we had no voice in the matter. Furthermore, the UWEs did not realize that with a pinch of courage they could have asserted the O script and thus greatly influenced the negotiations.

Midlife Crisis

After the workshop in October 1996 in Mlimani it was our job to start reorganizing the waterworks in the direction of water authorities operated as profit centers. We launched this part of the project in Jamala. In dialogue with the employees of the waterworks we had worked out a corresponding catalog of measures by December 1996. This essentially encompassed five areas: (1) a new organizational structure with a revised division of labor between the redefined departments; (2) a catalog of tasks for each position with the corresponding requirements to be satisfied by the person holding that position; (3) an incentive pay plan (calculated from a business standpoint); (4) a proposal for a new in-house works council; and (5) a list of positions that could not be filled internally. The latest possible time to introduce the new structures was set at July 1997, the first month of the fiscal year, at the end of which (in July 1998) the project would come to an end.

Together with the UWE from Jamala, I was able to present our catalog of measures to the Jamala waterworks board on January 4, 1997. One thing that was immediately obvious was that the session was conducted in the national language—which I as the key figure of the meeting did not speak—although all the participants spoke adequate English. Within the first few minutes the session was then declared a seminar, which meant it was not supposed to be a regular business meeting of the board at which decisions would be made. Finally, the UWE introduced my presentation in English, referring to it as an S&P proposal. In the follow-up discussion someone immediately asked what was wrong with the existing arrangement. They wanted to hear from me why any reorganization was being planned at all, and I was asked to conduct a corresponding study and present it to the board. This, of course, was where we had been before the feasibility study of 1994.

A main part of the discussion at the seminar dealt with the proposal for an incentive pay plan. The suggestion evidently triggered a fundamental anxiety. People expressed reservations, wondering if it was even compatible with Ruritanian legislation and regulations, and it was suggested that I should best check that out. The background to these misgivings was, first of all, a question of fairness. Although even the lowest-income groups would earn considerably more than they had previously, the board raised the question of whether it would be justified that the management earned so much more.

I commented that the management would only earn a high income if the operating income were sufficiently high, but that was evidently not considered a satisfactory response to the fairness question. The other reason for reservations was related to public opinion. What would people in the city say if they read in the newspaper that the waterworks employees would be paying themselves such high sums toward the end of the year? Finally, it was intimated that it would be better if the entire package of measures were first discussed calmly and in detail internally at the waterworks with all the employees at all levels of the hierarchy. Later someone said that the board was opposed to hiring new skilled staff and had guaranteed current employees' jobs. No minutes of the session were ever written up.

The seminar of January 4, 1997, gave rise to profound skepticism in the Jamala waterworks regarding the entire endeavor, a skepticism that previously had been latent at most. For example, the proposal contained suggestions for health insurance. This had been done under the assumption that the salaried employees would terminate their positions in the civil service and sign private employment agreements with the new "water authority." In such a case a number of aspects of their social benefits would have had to be restructured. We wanted to outline this even though there was no corresponding bill, in order to plan the financial implications in advance and thereby reduce the uncertainty for employees. In order to understand this point correctly, one should know that Ruritanian's public health system, which is available at no charge to civil servants, has totally collapsed. Everyone avoids going to a public hospital or polyclinic as best they can, since the service there has dropped to an unacceptable level. Doctors in private practice meanwhile receive hefty payments in cash. This is a constant subject of discussion among the workforce. To offset this untenable situation we proposed introducing a company health insurance fund, but ultimately this unsettled people even more.

In June 1997 the board nevertheless accepted a slightly revised version of the catalog of measures. Yet when the new fiscal year—and thus the final project year—began on July 1, 1997, the organization had not been restructured, the people had not been transferred, and there were also no new skilled staff and no incentive pay plan. Moreover, the board had not yet written to the undersecretary to obtain the necessary special authorization for these changes, nor had the UWE from Jamala made any effort to solicit such a letter.

The aforementioned bill supplement was in fact approved by the parliament in February 1997. But until today, July 1997, it has yet to be published in the corresponding gazette and consequently is still not in force. This is, however, hardly a surprise, since an amendment to the Waterworks Ordinance makes no legal sense in and of itself, as we have been aware for quite some time. In Baharini as well, this matter could no longer be ignored after February 1997. In accordance with normal procedures, the Water Ministry set out to draw up an Implementation Provision, to serve as a bridge from the amended Waterworks Ordinance to the practical work of the waterworks. The task force assigned to this project—which occasionally also called upon the UWEs in Baridi, Mlimani, and Jamala—found itself facing the same legal loophole that our project had been caught in since March 1996. The implementation provision—called the Waterworks Regulation—was supposed to regulate the actual implementation of the amended Waterworks Ordinance. But the most important part of this regulation, that is, the legal form of the waterworks, had not been laid down in the Waterworks Ordinance. So it was impossible to draw up an implementation provision until all the other relevant laws were also amended.

A conspiracy theorist would presumably say at this point: The Water Ministry was opposed to the privatization of the municipal waterworks from the very outset, since its influence and significance would then be reduced considerably. This cleverly precipitated predicament provided the ministry with a situation in which it could announce: "See, we did everything within our power. Now the autonomy won't work because the other ministries have not accomplished their tasks." From the conspiracy standpoint it looked as if the representatives of the water ministry had been blessed with a clever coup in fall 1996 at the workshop in Mlimani. They skillfully tricked the NDB into thinking that the amendment to the Waterworks Ordinance would cover all aspects of the intended reform. However, I suspect instead that conflicting interests and contingent shortsightedness brought about a result that no one had expected.

In any case, we were all stuck in the dilemma. The fourth quarterly report, which presented the project's progress from April to June 1997 and marked the halfway point of the project, served as the midway review that we recently sent to our clients (see figure 2.4 showing a sample page of this report as an illustration). In it we mentioned the three main obstacles facing the project: (1) Because of the way the NDB tied the technical program (TP) to our organizational improvement program (OIP), the technical water losses in the first year were not dealt with at all. Service quality thus could not be improved and the desired positive effect on the customers' payment morale never set in. Above all, however, the élan was lost as soon as it turned out that our holistic approach was not that holistic after all. Everyone started feeling that all we could do was organize the paperwork a bit and set up a couple computers, while the real engineering project was yet to come. (2) The continuing delay in determining the legal and political framework and the lack of alternative solutions led to our inability to complete very much within the first year in the area of organizational improvement. (3) The fact that we still did not have customer data—which were supposed to be supplied by the executing agents—meant that we could not activate the completed software in the course of the fourth quarter. Thus the most effective lever of the project was still not available at the beginning of the second year.

The bottom line for us was that we could not reach our goals because the necessary prerequisites were lacking. Thus through no fault of our own we would lose the 10 percent of the total personnel costs for the project that were withheld and would be unable to make a profit. Our only chance was to pull the emergency brake. I announced that, in accordance with the contractual conditions, I would terminate the consulting contract three months after this notice if the project-executing agencies did not resolve their problems concerning the general framework and the customer data. Obviously, I was interested in holding onto the project as long as any realistic prospects of success remained. I therefore proposed an approach to solving the problem: To rescue the project objectives, I supported extending the project running time by one year, until July 1999. I would accept such a regulation only under the condition that an adequate general framework would enter into force before July 1998 and that the project executing agents would supply us with the customer data within a negotiated period of time.

In my approach I deliberately did not go into detail regarding the matter of the missing customer data. In March 1996 the UWEs had claimed that they had already completed this task or that they would take care of it on their own. At that time we could not afford to alienate them and let the matter rest. I was well aware that the only solution was for us to revise the customer data ourselves. It must meanwhile also have become clear to the UWEs that as a result of the interface between software (on our side) and data (on theirs), in the end they would have nothing to show if there were no adequate data. They must also have become aware that we were simply waiting to finally receive a proposal from them to transfer the responsibility for the customer data to us, so they could avoid losing face.

End of Report by Julius C. Shilling

DOUBT

3 The Disclosures of Science

Preliminary Remarks : Edward B. Drotlevski

Although I probably should have anticipated it, I was nevertheless unpleasantly surprised when I learned from Shilling in July that there was already an anthropologist in the project. It is no longer possible these days to have "a field of one's own," even when one has carved out a relatively obscure area of expertise. At any rate, this offered me an opportunity to observe another anthropological observer in the field. The idea appealed to me and I decided to pay a visit to Dr. Samuel A. Martonosi, who works at the Urbania Institute for Social Research and is simultaneously a freelance project anthropologist for the S&P consulting firm. Coming from my somewhat rundown and chaotic university, I felt a little out of place at the luxuriously endowed Institute for Social Research, which is dedicated to the central questions of modern social order. As I did with my first two interlocutors, I will let Martonosi speak for himself here.

NARRATOR : SAMUEL A. MARTONOSI
Location : Urbania, Normland
Date : August 15, 1997

Positioning in a Field

In contrast to collectives lacking formal organization, businesses and bureaucracies are able regulate their access to outsiders with relative ease. Thus the first question for scholars engaging in empirical organization studies is how to obtain field access that is both reliable over time and does not obstruct their own research. The second question is connected to the fact that this kind of research involves "studying up", that is, as a researcher one is dealing with people who have equal or superior status in every relevant respect. These people are in a position to determine what kind of research into their own work they would like or will tolerate and how that work should be subsequently depicted.

The fact that the following three approaches predominate in the literature on development cooperation is related to this basic constellation of studying up. The first of these approaches is primarily concerned with the economic, political, and sociocultural structures of those societies that, according to general opinion, are supposed to "develop" or "be transformed." This kind of research examines the presuppositions, impediments, and consequences of development at the sites where development intervention occurs. The second approach is primarily concerned with conceptual questions: What kind of society should be offered as an ideal model? What should the much lauded civil society look like in detail? The third approach focuses on overarching questions such as the globalization of markets or the relation between culture, politics, and economy. Practical and everyday development occurs in the space between these three external poles. Such practices are rarely investigated themselves because it is presumed that they can be deduced from the three poles and are thus not an object of study *sui generis*.

The organizations involved do precisely what all organizations have to do: They present themselves to the outside world as if they were black boxes, in which nothing occurs except the orderly and rational processing of inputs into outputs.[1] If organizations do not produce the effects they are supposed to according to their own mandate, the metaphor of the black box suggests that we should turn our attention to the issues defined by the three aforementioned approaches, all of which lie outside the black box. If, however, we nevertheless want to understand how development occurs on a practical and everyday level, then it is necessary to find out what actually occurs inside the black box. According to the main rule of access—"No admittance except on business!"—people who want to see how the game is played from the inside have to play the game themselves. The "real" players must be able to count on the test players following the rules just as they do themselves. Otherwise they would not initiate these test players into the game. The best assurance exists when test players participate in the game with a dedication and commitment comparable to that of real players. One of the most important rules that test players must respect concerns the depiction of the organization to the outside world. This rule states that nothing noteworthy or remarkable actually occurs inside the black box. It is all merely procedures, files, lists, and tables—nothing that anyone has to see or experience first-hand in order to understand. Reference books are sufficient to find out exactly how it is done. Anyone who, once initiated, still wants to report about events within the black box runs the risk of playing the spoilsport.

This is the sociopolitical and ideological context in which development research takes place. The dominant players in the field—that is, those who distribute the money—simply pronounce with unquestioned certainty that anthropologists are responsible for the first of the three aforementioned issues, that is, for the social structure, culture, and local knowledge of those social worlds that are supposed to be developed. The discipline of anthropology also presumes that this will be the case because the core business of anthropology is to give a voice to the voiceless, and these are supposedly the same people who are to be developed or, inversely, protected from development. If anthropologists engage in this game unreflectively, they continue to operate within an established frame of reference that prescribes a particular perspective and a particular conceptual and interpretive schema. Borrowing from Luhmann, we could say that a social science operating in this way juxtaposes itself with the experts in the field and looks together with them at the same objects, that is, at the problems of so-called developing countries, and thus engages in first-order observation. A fruitful alternative to this would be to reflectively thematize this configuration in anthropological development research by including those translocal mechanisms usually classified as codified law, formal organization, science, and technology.[2]

The Organizational Model

From the perspective of a first-order observer, the problem definition here is unambiguous: On the one hand, there are rational models; on the other hand, these rational models are supposed to be adopted by culturally oriented human beings. If this process does not function properly and if, at the same time, the knowledge of engineers and management experts has been exhausted, then sociocultural experts will be called in for "postintervention criticism." As long as their diagnoses are restricted to the society and culture of the recipient of development aid, these sociocultural experts invariably provide arguments for continuing the process, albeit more cautiously. This well-rehearsed approach, however, is not as easy to criticize as it might seem, either on the basis of my contextualization here or on the testimony of those who regard themselves as advocates of local knowledge. After all, modernization theory provides background support. Despite decades of doubt and critique, the questions raised by classical modernization theory continue to have burning relevance—unfortunately also in the literal sense—and yet remain unanswered. In what follows, I will sketch in broad strokes the panorama that appears from this vantage point.

Today's complex societies of Africa are divided into social worlds that are perceived internally as consensus communities. In these separate social worlds, economic and political action is embedded socially and culturally, which means that economic and political rationality remain largely subordinated to social rationality. In the language of Max Weber, we could say that these social worlds view themselves as internal worlds that delimit themselves from their outside worlds by distinguishing between internal and external morality. Although the obligation to provide mutual assistance is highly valued in the internal world, it has practically no value in the external world. One consequence of this is that the social space between internal worlds becomes a moral no-man's land. Unrestrained acquisitiveness prevails here and an expansive venality develops among everyone who operates in this domain.[3]

The empirical casuistry derived by means of modernization theory allows us to identify the following basic pattern in the complex societies of Africa: Transactions that take place within the framework of markets or bureaucracies are secured not by trust in the system but primarily by personal trust. Wherever the ethics of markets or bureaucracies ought to prevail, the actors turn instead to the ethics of their internal world. In other words, the obligation of mutual assistance for real people is considered more important than the commitment to impersonal procedural rules. Protection and loyalty obligations within the framework of family, friendship, and neighborhood as well as that of patron–client relations are more important and more binding than compliance with impersonal rules, that is, rules that operate without regard for the person. As soon as these obligations are invoked, recourse to fraternalism increases recursively and becomes endemic. If two people have agreed to a transaction that violates the official rules of procedure, they will stick together through thick and thin in order to minimize the mutual risk. The result of this kind of association is frequently a particular form of anomie.[4]

Specific to institutions in civil society is the fact that these institutions—in contrast to personal obligations—are valid for people who do not know each other. This means that the obligation of players must be transferred from specific persons to the rules of the game. Only in this way can market and legal orders emerge that allow for reliably calculable behavior on the part of participating players who do not know each other or at least act as if they do not. Commitment to the rules of the game, however, presupposes that rationality is not subordinated to a consensus community of mechanical solidarity, because this would prevent strangers from coming together as a society. Instead, commitment to the rules demands a specific interpenetration (a distasteful term coined by Parsons) of economic, political, cultural, and communal rationalities. No society—as an aggregation of strangers—appears to be able to circumvent this condition. In my opinion, this is the probing question that modernization theory has raised, and none of its critics who are so fond of pointing to the different paths of modernization have been able to offer a better alternative.[5]

This question arises for development cooperation in a concrete way. The organizations that are well suited to act as the executing agency of development projects or that have been created specifically for that purpose are, according to their definition and mandate, located precisely in those precarious arenas where trust in the system is lacking. The executing organizations of development cooperation have the official function of institutionalizing the rules of the market economy, democracy, or rational bureaucracies. Accordingly their work must involve redefining the boundaries between internal and external worlds. They must contribute to the expansion of the rules of the internal world, in order to provide a moral "lining" to rational economics and administration without, however, imposing an extraneous logic on that world. In order for project-executing agencies to assume this function they must themselves already be organizations of this desired type, and the services they provide must be regarded as public goods, which all the country's citizens—regardless of their affiliation to an internal world—recognize as somehow useful for themselves. In reality, however, the arena of development cooperation in particular is predestined to fall under the aegis of external morality. It is precisely here that every possible informal network is connected in order to channel resources into the various internal worlds.

The Hidden Side of Organizing

The problems that become evident through the lens of modernization theory—that is, those of creating predictability in Africa's complex societies—of course look rather different if one changes the theoretical lens. One point in particular is important here: Every perception of a problem presupposes the vision of a better alternative. Only on the basis of this vision is it possible to even recognize something as a problem requiring a systematic solution.

And only on the basis of this vision do the contours of a particular evaluation and a particular approach to this problem as well as the possible solutions become evident. In the discourse of development cooperation, the following mechanism kicks in at this point: In the course of analyzing the problem, Western society and its organizational forms are transformed into ideal traveling models that provide the basis for identifying problems in developing countries. Over time, what had initially been merely a figure of thought becomes an objectivist assumption. In the discourse of development cooperation, idealized models from Europe and the United States—for example, civil society, the market, and rational bureaucracies—are ultimately regarded as tangible realities.

This leads subliminally to the impression that the perversions of Western welfare states brought about by their own political elites have nothing to do with the kleptocracies in Africa. As a result, an opportunity is missed to analyze the logic of the internal erosion of Western welfare states as refined and highly developed variants of the same structural problems that are more blatantly evident in the erosion of bureaucracies in the countries of sub-Saharan Africa. This also allows a key problem of capitalism to be projected onto the poor countries of the South. The social integration of industrial capitalism is predicated on the preposterous demand that we are obligated to its rules, even though these rules are organized in such a way that players are systematically rendered superfluous and replaced with more efficient players. Inclusion is always conditional and actually aims at exclusion. Inherent in capitalism is an explosive tension between competition and solidarity or between reciprocal and distributive justice, for which only extremely precarious solutions exist. Even in the countries of Europe and the United States, where capitalism has been practiced for more than a century, where it is firmly institutionalized and supported by religion and cosmology, the system requires perpetual growth. Only as long as there is growth and a sufficient number of players who have the prospect of improving their life situations— as long as the "pay-off coffers" are sufficiently full—is it possible to maintain the commitment to a set of rules that systematically renders players superfluous.[6]

In the development discourse informed by modernization theory, this unresolved central problematic of Western-capitalist social order is displaced onto the poor countries of the South, where it is in fact much more blatant and brutal. In sub-Saharan Africa there has been no real growth since the end of colonialism, and for this reason the official pay-off coffers are empty, allowing endemic forms of rule violation to emerge. As soon as the structural background of these phenomena is forgotten, the Western model of society no longer appears to be the cause of the existing problems but rather the solution. That which requires explanation is no longer the enigma of commitment to rules of exclusionary inclusion that are in principle directed against those who follow them, but instead the strategies of those who reject these paradoxical rules and thus retain established loyalties. If this reversal is no longer recognized as such, the unidirectional gaze at those societies that are supposed to develop appears to be the only logical perspective. If, however, attention is directed at what actually occurs within the black box where development is made, a completely different picture emerges.

Institutional Organizations

The first thing that a neophyte stumbles across in approaching the Normesian organizational field of development cooperation is the distinction between technical and financial cooperation, and with this the existence of two separate implementation organizations, the AOD (Agency for Overseas Development) and the NDB (Normesian Development Bank). This is particularly confusing because technical cooperation (TC) involves much talk of people and organizations, whereas financial cooperation (FC), in contrast, focuses more on technology. It becomes clear at some point that these terms refer to two different procedures. One still has to wonder, however, why the two should be separate from one another, since the distinction makes absolutely no sense in terms of the issue itself, as even locals in the field will confirm.

On closer examination, it becomes evident that the distinction between TC and FC can be traced back to budgetary rules of parliament. The urgently needed adaptation of these rules to practical issues in the field, however, is connected to a myriad of issues that are themselves interwoven with quite different and overarching procedures and political processes that everyone is loath to disturb because this kind of procedural modification always has unpredictable consequences. After all, the current regulations have been firmly institutionalized and are embedded in a power game. The AOD is a state-owned enterprise, in which the ministry for development cooperation (MDC) represents the owner. At the same time, this representative of the owner is *ex officio* the chairperson of the AOD supervisory board. That is to say, the MDC holds sway at the AOD. In the case of the NDB, which is also state owned, the ministry of finance represents the owner on the bank's administrative board, which means that the finance minister has the say here. Integrating the AOD into the NDB would therefore reduce the development ministry's role; and conversely, transferring the NDB's development departments into the AOD would reduce the finance ministry's role.

The distinction between financial and technical cooperation, therefore, is tied to established, power-saturated configurations, and the possibilities for reform are in turn dependent on these configurations. Practical issues that come up in specific projects—for instance, the yearlong delay of our project in Ruritania—are rather insignificant at this level. We find a virtually identical configuration in Baharini between the prime minister's office, the finance ministry, and the water ministry. Since 1992 they have been having difficulties granting the waterworks in Baridi, Mlimani, and Jamala more autonomy than proved possible by exhuming the almost forgotten Revolving Fund Act of 1965. More extensive and more modern changes would have seriously disturbed the established institutional order and the entrenched ministerial power relations.[7]

A second aspect that a neophyte immediately notices upon entering the Normesian organizational field of development cooperation is the banking rhetoric of the NDB, which is the main focus here. No one—for example, stockholders—entrusts the NDB with their own money in order to maximize returns, nor does the NDB seek out its borrowers according to their productivity or their securities, as a commercial bank would. The NDB operates not in the marketplace, but rather explicitly and intentionally where the market has failed. Nevertheless, in negotiations with African project-executing agencies and political authorities, bank representatives argue as if they were concerned with capital growth as determined by the laws of the market. I have often experienced this myself and observed how British colleagues from the Overseas Development Agency (ODA) and U.S. colleagues from the United States Agency for International Development (USAID) sneered at the banking rhetoric employed by the NDB staff, especially since African negotiating partners were never taken in by this rhetoric.

As an implementation organization, the NDB—contrary to its own rhetoric—is a typical example of what Meyer and Rowan call an "institutional organization": an organization whose success depends less on its results than on its adaptation to institutionalized expectations and its observance of procedural rules. This basic configuration means that the NDB encounters completely different problems than a profit-oriented enterprise would. The development bank's main problem consists in navigating between two irresolvable inconsistencies that the organizational environment expects of it.[8]

The first inconsistency lies in treating effectiveness and legitimacy as coequal demands. On the one hand, the development bank is supposed to be effective and efficient: With the means placed at its disposal, it is supposed to set in motion a maximum of sustained development. On the other hand, however, because it is an administrative bureaucracy, the NDB must also be particularly strict about maintaining its own guidelines, which are listed in a thick handbook and repeatedly cited in-house. These two goals, effectiveness and procedural fidelity, will continually conflict. Over time, our project has produced a long list of this kind of dilemma. To mention only the most crass example here, from the standpoint of effectiveness, the project could have started either in mid-1994 or else after the waterworks were granted autonomy. According to this logic, we should have aborted the project at the latest at the Mlimani workshop in October 1996. However, precisely such considerations of effectiveness and efficiency are not paramount for an administrative bureaucracy. Adherence to procedure is a principle that cannot be violated at any price. As a state-owned bank responsible for official development assistance in the realm of financial cooperation (FC), the NDB cannot, for instance, finance a project that was not evaluated according to a particular procedure agreed upon with the OECD, the Normesian MDC, and the government of the recipient country. Meyer and Rowan have identified this type of rationality: "A sick worker must be treated by the doctor using accepted medical procedures; whether the worker is treated effectively is less important."[9]

The second inconsistency that the NDB has to deal with is doing justice to criteria of both progress and emancipation. The meaning and thus the legitimacy of development cooperation are derived from the fact that it is recognized as the implementation of the metanarrative of the progress of human society. This grand narrative contains two dimensions that have tended to diverge in the process of its historical unfolding. On the one hand, there is the idea of progress as the technical modernization of the world, as the increase in material prosperity and the control of nature, as the liberation of science from extrascientific considerations, and as an ever-increasing and ever-accelerating process. In the field of development cooperation, an attempt is made to connect to this meaning-endowing and legitimating narrative of progress by identifying a corresponding development deficit in the countries of the South and promising the elimination of this deficit in the name of the universal progress of humanity. This is supposed to occur by transferring something from here to there in order to initiate sustained, belated development. Were this kind of deficit not assumed, the NDB's mandate would be meaningless. The development bank views itself as an instrumentally rational authority that is responsible for providing loans, technology, and expertise in order to overcome this deficit. Among the initiated, this is called a "technical fix."

The original version of the metanarrative of progress, however, aims not only for emancipation from the strictures of nature by means of scientific-technical development, but to an equal degree for social emancipation from the bonds of an unfree society. This means equal participation in the political decision-making process. In connection with this topos of the metanarrative, the postcolonial ideas of inviolable self-determination and of the unconditional equality of all nations and all cultures have come to be accepted in the relevant arenas. In order to appear legitimate and meaningful, development cooperation has to demonstrate that it is also the practical implementation of the emancipation topos of the grand narrative.

This means that the progress topos of the narrative is translated in development discourse into deficit elimination and thus into an existing inequality, while the emancipation topos of the narrative is translated into sovereign self-determination and thus into an existing equality. To satisfy these contradictory demands, the NDB falls back on the repertoire of institutional organizations: It concentrates on loose coupling, emphasizes its trustworthiness and integrity, and monitors performance as a ritual evaluation.

Securing Success

Loose Coupling   A widely recognized ideal presumes that an organization functions best when it does exactly what it is supposed to do according to its official external representation. In other words, the more tightly the organization's practices are coupled with this external representation and the formal structure of command, the greater the determinism and the better the performance. While this does not hold even for businesses that are primarily oriented toward increasing economic efficiency, it is utterly misleading with respect to institutionalized organizations. In order to exist, institutionalized organizations must always serve first and foremost the legitimation narratives assigned to them by their environment. However, because these narratives are always multiple and contradictory, the only way such organizations can survive is to couple the translations of these narratives in official representations (organization aims and structures) only loosely with the actual organizational practices. One of the specific implications for the NDB is that the topos of emancipation is articulated in the O script (the official script), while on the quiet, according to the U script (the unofficial script), it is more concerned with the topos of progress, which is addressed through deficit elimination.[10]

The principle of loose coupling generates a series of typical organizational symptoms. Institutional organizations celebrate their own trustworthiness and integrity and that of their individual staff as their highest values. While this phenomenon is obvious at universities, it can also be found in diverse variations in all institutional organizations. The NDB, I believe, offers a case study in this category that could rival any university.[11] Institutional organizations also attach great value to the professional competence of their staff and the subordinate institutional organizations they commission to execute assignments.[12] In our case, this symptom was already present in the MDC. The ministry, which is responsible for the political side of development cooperation, emphasizes that development cooperation is a very complex activity and that the ministry itself does not possess all of the required know-how. It does know, however, that it can confidently rely on the NDB, a highly professionalized implementation organization, to provide such expertise. This argument reduces the ministry's duty to supervise the implementation organization to purely technical dimensions.

Specifically this amounts to reducing the fundamental evaluation of the overall sustainability of development cooperation to the evaluation of individual projects. Just as hospitals do not produce "general health," but instead provide appropriate treatment to individual patients, and universities do not produce "knowledge," but award individual diplomas, development banks do not produce development, but rather initiate individual projects.

The integration of projects into overarching programs, in contrast, is omitted because this could easily lead to questioning the significance of the programs. Our waterworks offer a paradigmatic example of this widespread symptom in institutional organizations. In the beginning, a detailed coordination of the technical project (TP) with the organizational improvement program (OIP) was a core element of a comprehensive program to reform the three Ruritanian waterworks. Over the course of the project, however, this coordination dimension was abandoned. While in reality this separation raises fundamental doubts about both projects, this issue has been ignored at the NDB, where attention today is instead focused on a separate evaluation that can present a more positive picture.

The mechanism of loose coupling functions in the following way: The general aims of an institutionalized organization cannot be spelled out in concrete individual responsibilities because these are inconsistent and would result in contradictions. The NDB, for example, cannot simultaneously be as effective and efficient as a commercial enterprise, as bureaucratically correct as an administrative body, and as politically correct as a voluntary association of like-minded individuals. Loose coupling is the necessary pre-supposition in order to continue on a practical level despite these contradictory objectives and to arrive at a more or less acceptable result. This allows for maneuvering room, in which the prerequisites for developing trust and integrity as models of orientation can be established. These models of orientation ensure conversely that the complete decoupling of representations and practices does not occur, that is, that there are no "Potemkin village facades" disguising a completely different reality. They ensure instead that the external representations of these practices establish what Meyer and Rowan call "ceremonial facades." Like rituals and myths, ceremonial facades contain not false messages but rather condensed ones about the meaning of practices, without specifying their actual operations. Thus loose coupling and ceremonial representation for the outside world are not signs of the failure of institutional organizations, but instead indispensable means for ensuring success.

The fundamental dependence of an organization on loose coupling and the trustworthiness of its staff of course also remains the source of very specific blunders. The NDB staff, for instance, often find themselves in the uncomfortable position that a consultant, whom they have contracted, can easily deceive them owing to their enormous spatial and social distance from the reality about which the consultant is reporting. However, the uneasiness of the NDB staff results, above all, from the fact that in development cooperation there is usually no table of relevant and stable data for orientation that they could use to check the plausibility of the consultant's report. They continuously operate without any firm ground beneath their feet. Instead of responding with caution and modesty, the NDB staff tend to suspect their consultants of invoking deceptive smoke and mirrors.

This somewhat strange inversion is also related to the fact that the NDB does not commission most of its expert reports to consultants because it actually requires more or different information. The usual motivation for such contracts is instead the fact that a better decision can often be made between already existing options if an external opinion is brought in that could tip the scales. There are two advantages to this strategy. First, it is easier to diffuse internal differences of opinion and growing conflict if at least some of the responsibility for a decision can be shifted to a party outside the institution. Second, this shift can also prove advantageous if problems subsequently arise, since mistakes can then be attributed to the external expert. For the NDB staff, this kind of outsourcing of problems and failure avoidance is part and parcel of the prevailing norms of integrity and professionalism.

It is also the case, however, that a vicious circle results. These procedures are not only caused by but also actively promote a sense of infallibility among organization members, thereby engendering structural myopia and institutional amnesia. According to the O script, the NDB delegates the implementation of projects to executing agencies and consultants. In reality, however, projects are implemented primarily according to the U script—and thus according to the ideas of the NDB—although this cannot be stated officially. Nevertheless, the NDB staff begins at some point to believe their own official representations that have in fact been produced for external consumption. There is also an eminent motivation for this perception, since in this light— that is, according to the O script—they can never be held accountable for project failures.

Ritual Evaluation   Institutional organizations succeed—or rather they must succeed—in translating performance assessment into ritual evaluation.[13] According to official definitions, development cooperation is not a commercial enterprise and is therefore unable to establish any kind of price mechanism that would distinguish success from failure. The emphasis cannot be on commercial results, but is shifted instead onto success indicators and thus onto procedures that establish those indicators. Furthermore, the significance of procedural oversight is tied to the fact that the lender's relationship to the borrower is marked by habitual mistrust because borrowers in development cooperation are selected on the basis of their credit-unworthiness. Analogous to this, a habitualized mistrust also predominates with respect to consultants because their superior knowledge allows them to easily deceive financiers. As a result, a dense network of procedural rules and oversight mechanisms has developed. Officially this is supposed to ensure that the resources provided are used effectively and efficiently and exclusively for the purposes for which they were originally designated.

An evaluation procedure is agreed upon that will not call the system as a whole into question. Institutionalized organizations that are compelled to fulfill contradictory expectations have no choice but to proceed with evaluations in the same manner as they do with the loose coupling of formal structures with actual practices. Evaluations have to be organized in such a way that they do not reveal the inevitable inconsistencies of institutional organizations. In concrete terms this means the following. On the one hand, everyone knows that ultimately the only question that actually matters is: "Has the input provided by development cooperation initiated sustained development, leading to a better life?" On the other hand, the entire organizational field is constantly engaged in producing facts and figures that are supposed to measure the direct effects of individual projects in clearly defined contexts. Appropriate indicators for this must be defined from the outset so that precisely the central, decisive point has been factored out of the equation. This is in the interest not only of the directors of the NDB, but also the overarching institutions. The regulatory agencies—that is, the ministries of finance and development—have an interest in the NDB (which they supervise) being evaluated in such a way that no public embarrassment results. In this sense, I don't think it's astonishing at all that a staff member at the MDC said that he was an "NDB fan."

Both the necessities and the possibilities of ritualizing evaluation are greater in the realm of development cooperation than in other fields of institutional organizations, for instance that of public health or education in the donor countries, where the procedure is similar. The necessities are greater, simply because the contradiction between the imperatives of official representation arising from the narrative of equality and the practical instructions for action arising from the narrative of progress is more pronounced than comparable contradictions in other fields. The possibilities for ritualizing evaluations are also greater in comparison because the practical work of development cooperation takes place so far away from the donor country, in this case, in Ruritania. The otherwise common mechanisms of criticism and correction through the media or political channels apply either hesitatingly or not at all. The sociopolitical realm in which the achievements of development cooperation are realized is kept more or less completely separate from the sociopolitical realm in which decisions are made about the conception and evaluation of development cooperation.

In other words, those entities that are supposed to be monitored and evaluated—the institutional organizations that transfer resources to poor countries in the name of the taxpayers, in order to set sustained development in motion there—are simultaneously also the entities that possess a largely unchallenged representational monopoly on their own actions. And the principal, the MDC, which commissions the NDB as its agent to assume certain responsibilities in its name, has its own interest in maintaining this representational monopoly at least in regard to the public.

The bottom line is that the mechanisms described here for ensuring the success of institutional organizations lead to the conviction that development cooperation involves the transfer of preexistent models. Without this conviction—known in the jargon of the field as the "blueprint approach" and in science and technology studies as a "standardized package"—it would be impossible to maintain this ritualized safeguarding of success as practiced by leading institutional organizations. Without objectives of intervention that have been determined beforehand, it would be impossible to provide an account of the achieved success rates, and without accountability the legitimacy of development cooperation as a whole would be called into question in a radical and fundamental way. Although there is widespread and vehement critique of the blueprint approach even within the field of development cooperation, this is ultimately little more than lip service under the existing circumstances.[14]

The conviction that development cooperation involves a transfer of models has in particular one far-reaching consequence. If this intervention does not do what it is intended to do, the framework of the prevailing interpretive schema suggests the following explanation for this failure: The models were erected in the developing country—in this case Ruritania—merely as Potemkin villages in order to continue the usual game behind the facade. Even if we grant this diagnosis a certain empirical plausibility, it is nevertheless based on a fundamental error. The diagnosis that a facade without substance has been constructed contains an implicit assumption, namely, that the transfer of models could have been improved if there had only been a tighter coupling between the model and the local practices. This, however, conceals the fact that the success of the model in its original location—that is, Europe or the United States—was itself based not on tight but on loose coupling. Moreover, the fact that development cooperation itself can be organized only with loose coupling is completely lost in this interpretive schema.

This in turn ultimately leads to the fact that all error diagnoses made within the framework of this interpretive schema do not apply, because they divert attention from those mechanisms that in reality lead to the successful functioning of such organizations and from the actual mechanisms of development cooperation itself. This also obstructs insight into the errors of development cooperation. The sole problem officially recognized from this vantage point is located in the cultures of the recipient countries. This is where officially anthropology enters the development arena and, accordingly, its interests and competences are supposed to be limited to these societies or these particular kinds of problems. By complying with this expectation, anthropology provides an invaluable service to the self-staging of development cooperation.

The Epistemic Model

The institutional arrangement that disciplines the gaze of development anthropology according to the requirements of the organizational field and directs its focus to sociocultural factors is thus related to the established and power-saturated organizational mechanisms described above. Perhaps even more importantly, the institutional arrangement also implies an epistemological presupposition with analogous results. It is assumed that there are objective structures "out there in reality." It makes no difference whether the people operating according to this presumption speak with skeptical resignation about an iron cage, with revolutionary pathos about the colonization of life-worlds through system rationality, or with sanguine affirmation about human progress. Independent of their diverging political appraisals, the representatives of these different approaches all presume that behind observable developments, objectively existing forces are at work. And all three of these approaches claim that they stand outside of that which they are observing, describing, and practicing. In doing so, they remain within the basic trope of modernity: Scientific knowledge seeks to render the social and natural world controllable, to discover the inner laws of these worlds in order to use them to establish a better world.

All corresponding approaches to development cooperation are thus based on a shared distinction that they uncritically accept: They produce nomological knowledge that is then used to establish social structures in distant social worlds. When things do not work as predicted, the need for social-scientific explanations is shifted onto those local phenomena that do not accord with the pre-established laws. This distinction between norm and deviance means that on a metaphorical level it is not the cobblestones—that is, organizations, states, and markets—that require explanation, but only the grass growing through the cracks in the pavement. Some may regard this grass as bothersome weeds, while others see in it a green shimmer of hope for an authentic life-world beneath the stones. Both of these positions, however, take for granted that the cobblestones themselves have simply appeared self-evidently over the course of time and are thus hard as rock.

Verifiable findings about social constellations are compiled and the results are interpreted as the effects of the underlying causes. The figure of thought is always the same: This is the phenomenon and this underlies it. That which is presumed to underlie the phenomenon (for instance, social differentiation or/and rationalization) is regarded as the real force—the hard cobblestone—that exists independent of human observation and representation, even though it has just been conceived by the human mind. The things that the social sciences aimed to explain in the first place—the structures and laws of society and culture, that is, the cobblestones—are factored out in this kind of analysis. *Explanans* and *explanandum* are transposed. As long this kind of procedure appears to be self-evidently correct and objective, there is a reciprocal stabilization between hegemonic models of society and social analysis. This stabilization is based on a confusion of cause and effect.[15]

In the analysis of development and transformation processes, this reversal of cause and effect leads to stylizing the Western world—as the apex of system rationality—into the standard for human history. In contrast, all other societies appear to be deficient and in need of deliverance from the darkness of their own culture. Admittedly this result contradicts the primarily honorable intentions of those authors often concerned with emancipation, who would actually like to represent those human beings and life conditions who have no voice. However, it is precisely this fixation on the informal, the marginal, and the oppressed—that is, on the grass—that leads to the formal, the central, and the dominant—the cobblestones—drifting into an unattainable sphere of system rationality, which as a result remains beyond critical analysis and the sole alternative. One of the side effects of this distinction is that economics becomes the leading social science, responsible for the advance of system rationality under the category of "the market." One indication of this is that the predominance of economists in large development organizations is accepted as a given. The other disciplines are then automatically responsible for the consequences of this rationalization and for the impediments to it that intermittently block its path: the "informal," cultural, and so-called local knowledge. Descartes was the pioneer of this fundamental distinction, as Ernest Gellner argues in a pointed passage:

Descartes wishes, cognitively speaking, to be a self-made man. He is the Samuel Smiles of cognitive enterprise. Error is to be found in culture; and culture is a kind of systematic, communally induced error. It is of the essence of error that it is communally induced and historically accumulated. It is through community and history that we sink into error, and it is through solitary design and plan that we escape it. . . . The most important point about Descartes's account of the human condition is this: in order to avail oneself of reason, and to escape from culture, one must transcend all the errors which culture instills, and one must heed the inner compulsion of a special kind.[16]

There is almost no one today who would advocate the position described here in such an undiluted form. Nevertheless, I am convinced that it constitutes the firm ground upon which development cooperation takes place. According to Gellner's apt summary, this ground can be described by following two key aspects of rationalization.

There are two distinct but related elements of the rationalization of the world. On the one hand, there is the progressive and comprehensive assertion of the principle of efficiency, in Gellner's words, "the cool rational selection of the best available means to given, clearly formulated, and isolated ends." This principle is considered rational insofar as the sole admissible criterion for the selection of means is ideally the optimization of the means–ends relation. On the other hand, there is also the progressive and comprehensive assertion of the principle of coherence or symmetry: the like treatment of like cases. The principle of symmetry is an involuntary consequence of the principle of efficiency: Confronted with the same type of problem, an efficient agent will not only eliminate "irrelevant" considerations, but will always choose the same type of solution. In this way such agents are able to multiply their efficiency. Thus rationality here means treating similar cases in a similar way, maintaining regularity and coherence, and thereby creating a well-defined regime of action. In this way, actions become transparent, predictable, and controllable. The power of the control, however, becomes largely invisible insofar as it is synchronized with the internal laws of these processes.[7]

Although both sides of rationalization—efficiency and orderliness—are clearly important for economy, bureaucracy, and science, distinct nuances have emerged so that business organization has come to stand for efficiency, bureaucratic organization for coherence, and scientific organization for objectivity. The reason for this is that beyond a certain point attention to the orderliness principle reduces efficiency, just as conversely beyond a certain point following the principle of efficiency violates the principle of orderliness. We would ultimately not arrive at the truth if in the process we placed too much weight on costs and rigid principles of order.

In addition to these two elements—orderliness and efficiency—there is a third fundamental element of rationality underlying the other two. The correspondence theory of truth presumes that rational statements about external referents correspond to what those referents actually are. This gives rise to the premise that there is, to borrow from Gellner, "a common measure of fact, a universal conceptual currency"—which would mean that "all facts are located within a single, continuous logical space." And this in turn means that "all statements reporting them can be conjoined and generally related to each other." It also means that we can translate any random language into a single standardized idiom that refers to a coherent reality and that encompasses that reality.[18] In other words, it refers to a *metacode*, into which all *cultural codes* can be translated.

The principle of the separation of all separable, the *esprit d'analyse* as characterized by Descartes, is also part of the equalization and homogenization of facts. Concretely this means, for example, that the optimization of the means–ends relation requires that the scientist–bureaucrat–entrepreneur be able to separate the moral, aesthetic, economic, legal, and political dimensions. Rationalization in this sense is instrumental and tends toward the pervasive assertion of coherency and efficiency. According to the grand narrative of modernity (which provides meaning and legitimacy), progress is thus initially rationalization as cognitive progress. What modern human beings do not yet know today, they believe they will discover in the future. Like capital, knowledge in modernity appears to accumulate unceasingly.

Development cooperation, as it has been practiced for almost fifty years, makes sense only on the ground outlined here. However, for the past twenty or so years of development discourse this ground has been buried in a rhetoric of cultural relativism. According to a key distinction of modernity, epistemic relativism is rejected as a matter of course, while enormous efforts are expended to weave ethical and aesthetic relativism into the veil of development cooperation. I believe that the vicious cycle of errors, in which development cooperation is trapped, is integrally connected to the official restriction to the epistemological model outlined here. To find a possible way out, we need to adopt a perspective that will allow us to observe how organizational conditions and epistemological presuppositions in the field give rise to the rules of the game in the global organizational field of development cooperation. Only then can we turn to the rational objective of the practices of this field.

The Hidden Side of the Epistemic Model

The institutionalization of the rationalization idea is the only institutionalization that inevitably delegitimizes itself and thus deinstitutionalizes itself over time. It is this disillusionment of enlightenment that occurs before our very eyes while we debate nuclear fission, climate change, mad cow disease, biotechnology, and anthropotechnology, but also sustainable global development. We can distinguish three related mechanisms in this process of the inexorable self-delegitimation of rationalism.

One specific trait of postindustrial society is the fact that the character of labor has changed radically. Work, as Gellner points out, is no longer essentially the manipulation of things, but of meanings. Together with the industrial division of labor and the mobility of the workforce, this develop-ment demands a particular capacity of human beings: In order to make themselves understandable independent of context, they have to com-municate in a standard idiom. People who have learned a profession in India—for example, our three engineers at the waterworks in Baridi, Mli-mani, and Jamala—have to be able to work in Ruritania or in Normland and read the handbooks of their profession, which in part have not yet been written.[19]

Knowledge that can be communicated in the one admissible universal standard idiom must be regarded as objective and thus always as scientific knowledge, which, in contrast to narrative knowledge, exists independent of all social contexts. This kind of knowledge, however, does not arise in a vac-uum, but in separate and specialized institutions (I will return to this issue later when I discuss "centers of calculation"). The outsourcing of knowledge production behind the walls of such institutions in turn alters society's rela-tionship to its own knowledge, which it necessarily regards with increasing suspicion since there is no such thing as a scientific expert opinion without a counter expert opinion.

With this particular institutionalization of knowledge production, cogni-tive self-grounding has also become problematic, and this has triggered the second mechanism of self-delegitimation. The irresolvable enigma of how knowledge can be grounded in itself and what role ideology and power play in this process is presumably as old as humanity itself. This enigma, however, fully unfolds in the phase in which modern industrial societies (which reject transcendental authorities and invoke exclusively self-grounded knowledge) have transformed into postindustrial information societies. Information societies are the first societies to be conscious of their own communication as a reality and as a problem. Jean-François Lyotard is one of first astute analysts of this aporia:

With modern science, two new features appear in the problematic of legitimation. To begin with, it leaves behind the metaphysical search for a first proof or transcenden-tal authority as a response to the question: "How do you prove the proof?" or, more generally, "Who decides the conditions of truth?" It is recognized that the conditions of truth, in other words, the rules of the game of science, are immanent in that game, that they can only be established within the bonds of a debate that is already scientific in nature, and that there is no other proof that the rules are good than the consensus extended to them by the experts.[20]

The impossibility of proving objectivity is thus demonstrated by applying the scientific claim of objectivity to that claim itself. This alters the constitution of the rules of a rationalist proof procedure, since the claim that these rules are authorized only by a consensus of experts immediately raises the following question: Who decides what knowledge is, and who knows what needs to be decided? These questions lead in turn to the legal-political order of those who have to agree on the rules of the game and who require public authorization and recognition for this. Thus it is ultimately a legal-political settlement that assists us in distinguishing correct statements from incorrect ones. According to this understanding, objectivity is no more and no less than the conformity of a proposition to the grounding norms that are valid within the community authorized to determine whether the statement is correct or incorrect.[21]

The third mechanism that has led to the self-delegitimation of rationalism derives from the other strand of the metanarrative of modernity, that is, the emancipation narrative. This observation is also probably as old as humanity itself; however, only in the context of postcolonialism could it unfold in its current intensified form. From the perspective of the emancipation narrative, rational discourse appears to draw its legitimacy primarily from the autonomy and equality of the interlocutors. This autonomy and equality means that interlocutors are not required to subordinate their free will to truths that are regarded elsewhere as valid. This rule of the game is also legitimated by the impossibility of proving that a "true" statement necessarily leads to a normatively "good" statement and subsequently to a prescriptively "correct" statement. For this reason, the emancipation narrative requires a division of reason into cognitive-theoretical reason and practical reason. This division revokes the authority of scientific discourse oriented around the quest for truth to regulate practical life decisions. As a result, the relationship between truth, on the one hand, and justice and consensus or solidarity, on the other, becomes more complex. After the emancipation narrative has itself been emancipated from the rationalization narrative, both narratives continue to be juxtaposed as language games with their own respective rules. Insofar as the validity of the metanarrative of progress is contingent on the ability to integrate these two narratives, it has delegitimated itself by this separation.[22] Development cooperation, however, makes absolutely no sense without the belief in progress that brings about emancipation precisely through the ability to control social change. The question is how do we deal with this contradiction.

Translation Chains

Although the progress narrative has become rather tarnished in its native environs, it remains largely untouched in development cooperation. Even more astonishing, however, is the fact that worldviews that thrive in the South and radically challenge the modern narrative of progress are practically nonexistent in development discourse.[23] More precisely, they do exist but only in a tellingly reduced form—not as alternative dispositifs of reflection holding out prospects of a different world, but as sociocultural factors, whose effects are calculated into development discourse in order to better steer progress. Where does this peculiar blindness come from?

The official legitimation of development policy has to accommodate two mutually contradictory narratives. The narrative of the unequal distribution of progress and the subsequent duty of more developed nations to contribute to greater distributive justice has top priority. This, however, inevitably implies the designability of progress, for otherwise the duty of the more developed to come to the aid of the latecomers would make no sense at all. The idea that progress is designable necessarily implies the possibility of universal objectivity, or else designability would be inconceivable. The second priority in the official legitimation of development policy is the narrative of the equality and sovereignty of all cultures and all nations. This narrative is also concerned with justice, albeit from a different perspective. Justice through designable progress refers to a universalist worldview, whereas justice through sovereign equality refers to a relativist worldview. These two perspectives cannot in principle be valid simultaneously. Development cooperation resolves this problem by assigning them to different contexts.

Between idea and manifestation, this paradox traverses several sections of a longer *translation chain* that is interwoven with other chains into a network in which ideas and artifacts circulate on a global level.[24] The first section of this chain—even if it doesn't really have a beginning, our depiction has to start somewhere—links the Ministry for Development Cooperation (MDC) to the public, which enters in the form of the media, individual politicians, and political parties. One influential current of development-policy discourse calls for greater distributive justice in the world. To this end, it demands more progress in poorer countries through greater involvement by the government of Normland, which is supposed to allocate a larger percentage of its gross national product for this purpose. The professional expertise of specialists in the field, in contrast, indicates that developing countries have already exceeded their capacities for the productive incorporation of such input. Judging from the facts of the matter, it would actually be more appropriate to reduce the input. However, this argument can only be substantiated if the public is shown that programs fail precisely because these capacities have been exceeded.

Such evidence, however, would undermine the trustworthiness of the entire system that has developed between taxpayers and aid recipients to ensure proper transfer of resources. People who are convinced that their tax monies are well invested in development cooperation must be spared any insight into the contingencies and inconsistencies in the decision-making process. Ms. Hoff of Urbania, who sees part of her income tax used to provide Ms. Kimambo with better water service in Jamala, should not learn too much about the myriad of precipices that lie between her bank account and Ms. Kimambo's water faucet. Insight and the ensuing disillusionment could have serious consequences for the entire institutional arrangement. A disclosure of the actual problems could even serve to dismantle an entire worldview; as a result, well-meaning Europeans would have to resign themselves to the fact that some countries in sub-Saharan Africa have a life expectancy of fifty years and a child mortality rate of thirty percent because these countries are simply unable to process the aid that they receive. This kind of shock, however, cannot be incorporated into the political agenda and thus cannot be the goal of the political leadership in a ministry. Cautious announcements of the meager prospects of success would hardly alter the call for more aid. Such announcements would rather make people doubt the competence of those responsible for development cooperation. Hence, the only available option is to construct a ceremonial facade to hide all attempts to balance out the inherent contradictions as skillfully as possible.[25]

Of the many translation chains within the translation network, we can follow the chain leading from the ministry to the development bank. In transferring the paradoxical task of development cooperation from the MDC to the NDB, two other perspectives based on the logic of this institutional organization become increasingly apparent. This section of the translation chain is primarily concerned with procedural correctness and efficiency, which serves to strengthen and concretize the narrative of the rational designability of the world in a particular way. First and foremost, the implementation organization must ensure that it has convinced the MDC and the public that it has carried out its responsibilities correctly and has not spent more money than necessary. The relationship between procedural fidelity (as bureaucratic methodology) and efficiency (as goal orientation, which is less concerned with means than with results) is not only one of mutual conditionality, but at the same time one of inevitable and irresolvable contradiction. For the institutionalized sector of the organizational field—to the extent that this sector is located in the donor countries—this contradiction is defused, as explained earlier, by establishing ceremonial facades, loose coupling, an aura of unquestionable competence and trustworthiness and by ritualizing evaluation mechanisms.

From the NDB one chain leads directly to the Ruritanian waterworks; a second chain also leads to the Ruritanian project-executing agencies, but via the private consulting firm S&P. The two strands are then rejoined. The first chain—between financier and executing agency—places greater emphasis on the principle of equality. The partners of the development cooperation appear as project-executing agencies, which are provided with resources that they should in principle manage on their own. Thus we have followed the translation of the emancipation narrative to the O script, which has already been discussed previously. In contrast, within the second chain running parallel—between financier and project-executing agency via the contracted consultant—it is above all the principle of efficiency that comes into play. Here ceremonial evaluation has to be translated into commercial accountability, for otherwise the NDB would be decried for wasting taxpayers' money.

That which was still connected in the previous section of the chain—that is, between the MDC and the NDB—is separated here: In the one chain, the principle of equality (and with it relativism) pervades, and in the other chain, the principle of efficiency (and with it universalism). However, insofar as they both deal with one and the same issue, a trap has been built into the process. A second script is introduced to circumvent this trap: Contrary to the premises of the O script, the contractor, according to the U script, is surreptitiously placed directly under the supervision of the financier so that the latter can better monitor the efficiency of the former. In this way, the relativistic principle of equality, which predominates in the politically correct interaction between financier and executing agency, is not openly called in question through the universalist efficiency principle. A performance according to two scripts, however, is rather unpredictable. The frequent concern in the organizational field about the trustworthiness of negotiating partners and the widespread obsession with facts and figures are expressions of the fact that the equality principle and the efficiency principle can only be partially separated and that inconsistencies regularly arise when the two principles come together.

In the section of the translation chain in which the everyday and practical development cooperation between the consultant and project-executing agency finally occurs, a new circumvention of the contradiction between participation (relativism/particularism) and model transfer (objectivism/universalism) must be found that is appropriate in this context. The World Bank, which sets the tone in these matters, has officially stated that self-determination on the part of the African project partners is the most important prerequisite for the success of a measure.

It is the actions of the Bank's borrowers that will ultimately have the greatest impact on project quality. For this reason, stronger borrower commitment to, participation in, and *ownership of Bank-financed options* are essential to success. . . . The challenge for the Bank is to change the way it interacts with borrowers, from a pattern dominated by prescription, imposition, condition-setting, and decision-making to one characterized by explanation, demonstration, facilitation, and advice. Such a perspective . . . will lead to stronger borrower commitment and institutional capacities, better accountability for project performance, and ultimately a better Bank loan portfolio.[26]

This official statement neglects to mention that self-determination can easily conflict with the guidelines of the donors. Another aspect, however, is even more intriguing. When the World Bank demands that African partners should now finally assume "project ownership" or take the "driver's seat," it implies that the urge for self-determination is not simply a universal predisposition that inexorably exists but rather something that needs to be established. How could this be? Why shouldn't the borrowers have a natural interest in their own self-determination? Taking the World Bank's demand seriously, we have to ask ourselves where the problem lies in this section of the translation chain.

Representatives of the African administrative and managerial elite—in our case, the Ruritanian UWEs (urban water engineers)—confront their interlocutors, for example our S&P team (experts who have come here as adherents of the US–European progress narrative), with two plausible assumptions. First, they assume that we are convinced of the universal validity of our own models; otherwise we would not be considered experts and would not be so highly paid. Or to put this more generally, otherwise the World Bank, the IMF, the NDB, and other organizations of this kind would not place conditions on the allocation of loans within the framework of policy-based lending. After all, such conditions can only be justified if the people setting them are convinced that they are in possession of objective knowledge as well as the models derived from such knowledge. If not, what would be the sense of setting conditions and how could these even be defined in the absence of such models? Second, they presume that there must be at least a grain of truth in this conviction, for otherwise the West would not be as superior as it is.

However, when these highly paid Western experts now propose that discussions be conducted as dialogues because all truths are only relative and because the project is ultimately concerned with self-determination, their Ruritanian interlocutors presume that this is a tactical deception. Depending on the context, they suspect the motive to be a variety of interests, which can be reduced to a few themes. In the first place, they believe the talk of self-determination is simply a notorious form of domination disguised as participation. As the colonized recognize most readily, the true colonizer normally assumes this standpoint: "We are not content with negative obedience, nor even with the most abject submission. When finally you surrender to us, it must be of your own free will."[27]

The policy-based lending of development cooperation, which is associated with the catchword "participation," is a paradigmatic example of this. Our project is no exception in this regard. First it was vehemently asserted that the water sector had to be reconstructed, then a model of privatization was proposed, and then we called it "participation." The reaction to this power game is as classic as the colonization model itself: The Ruritanian project partners prefer the transfer of preexistent models, and they resist as best they can the intervention of consultants in translating this model into the local context. In this way, they attempt to remain in charge, at least with respect to this final point.

In some cases, it is insinuated that the participation demanded by consultants only serves to induce African interlocutors to divulge their hidden weaknesses, thereby enabling the consultants to pass the buck later on. For example, as consultants for the Ruritanian waterworks, we—that is, S&P—have repeatedly found ourselves, over the course of the quarterly reporting periods, billing more person-months than planned. In order to justify these person-months, we have to reach an agreement with the project-executing agencies about the necessity of this additional input. The more unexpected shortcomings and gaps that appear on the side of the executing agencies, the more input we can and must provide. An obvious insinuation is that behind our ostensible interest in participation is really the desire to identify the mistakes and shortcomings of the executing agency in order to use this as justification for increasing our input and thus our own profit. Our conflict with the UWEs about customer data also fits into this category. With this suspicion in the back of their minds, African interlocutors have little or no interest in participating in dialogue about developing locally adequate solutions.

The bottom line of the Ruritanian interpretation is thus that the official call for participation is really only a smoke screen for hegemonic dominance. This suspicion gives rise to the notorious defensive communication typically encountered in power-saturated negotiation processes.[28] Over the course of the project, which has been going for more than a year, almost all of the significant problems that have arisen can be explained in part as consequences of this power-distorted communication. During the inception phase of the project in March 1996, we employed every means possible as consultants to ensure the participation of the UWEs. Our efforts here were enormous, and Shilling was finally convinced that the inception report we submitted was a joint project. After the NDB issued its devastating critique of the report in a letter of May 1996, the three UWEs responded with an implicit clarification of the respective roles. By subscribing to the NDB's criticism, they shifted the authorship of the report onto the consultant with a truly bewildering lack of compunction. Another highlight was the workshop in Mlimani in October 1996. Whereas the official purpose of the workshop was to engage in dialogue about the best organizational form for the waterworks in Baridi, Mlimani, and Jamala, it proved impossible to conduct any kind of issue-related discussion at the workshop. The entire event focused on finding more or less diplomatic forms for the preexistent hegemonic assertion of the privatization model and for the preexistent rejection of this model through defensive communication.

The Technical Game and the Hegemony of Centers of Calculation

The translation chain from Ms. Hoff to Ms. Kimambo, which I examined in detail earlier, can be divided into several sections. In each of these sections, the basic paradox of development cooperation—that of being universalist and relativist at the same time—is processed according to the prevailing conditions of that section. Only one assumption is regarded as indisputably valid in all sections of the chain: that one can be a relativist toward everything except facts and figures. Conversely, we could also say that the translation chain consists of a sequence of mechanisms that serve to process objective facts and figures. On closer examination, this definition of the translation chain proves to be the elementary act of creating hegemony. If facts and figures are ascribed universal self-evidence because they are ostensibly grounded solely in the objects themselves, the inevitable dependence of that self-evidence on frames of reference and procedures is rendered invisible. Procedures that are embedded in a web of beliefs as well as a conceptual and a legal-political framework are in this way saturated with power. They appear to have an objective foundation and can be disseminated universally under this guise without further review. Players in the field call this a "technical fix."

Following Lyotard, we can distinguish between a *denotative* game, which is concerned with knowledge and thus with the true–false criterion; a *prescriptive* game, which is concerned with normative issues and thus with the just–unjust criterion; and a *technical* game, which is concerned with improvement and thus with the efficient–inefficient criterion. A technical move is "good" if it "does it better" than another move does and/or if it "consumes less" in achieving the same end. With only one restriction, development practitioners mean precisely such a "technical game" when they use the expression "technical fix": namely, a measure in which the efficient–inefficient criterion can be substituted apparently without consequences for the true–false, just–unjust, and beautiful–ugly criteria. In doing so, they act as if the technical game is appropriate because the normative dimensions are only of secondary importance. This position, in other words, says that the objective facts have to be determined first and this can be accomplished only independently of normative considerations. Sometimes it is necessary to include issues of normativity and aesthetics at the end of the process as sociocultural factors.[29]

In contrast to the epistemic community of development practitioners, Lyotard uses the technical game to point to the foundational problem of human knowledge, a problem that official development discourse wants to avoid at all costs. This foundational problem is the result of the aforementioned rupture between skepticism/knowledge, on the one hand, and evaluation/prescription, on the other. In other words, the foundational problem arises from the distinction between the dispositif of knowledge and the dispositif of emancipation. The separation of these two dispositifs is the very condition for postulating a denotative language game with universal validity. For knowledge to be accepted as "true," its independence from human desire first has to be acknowledged. However, this separation also means, conversely, that knowledge that has been accepted as true can have no binding consequences for normative issues. The objectivity of a statement, therefore, can be secured only through its separation from the dispositif of emancipation, which also means that it is devoid of any political significance. Transferred onto the domain of development cooperation, this means the following: If the technical game claims universal validity, it is for this very reason irrelevant to the dispositif of emancipation. If, however, the technical game is supplemented with a justification drawn from the dispositif of emancipation, it then loses its claim to universality. Both of these variants imply that development cooperation is impossible as a purely technical game.

To put this somewhat differently, normative and aesthetic dimensions are inextricably interwoven into the fabric of the technical game from the very beginning. At the same time, normativity and aesthetics have merely relative validity, that is, they are valid only within a particular framework. This relativism, however, also undermines the very possibility of the technical game and must therefore be rendered invisible. To the extent that such invisibilization is successful, the unavoidable relativism of the technical game can be passed off as universalism. If a social practice is understood in this sense as a technical game, then the connection between knowledge and power is rendered invisible, and the political parameters of the game can be established all the more effectively. Thus the official conceptualization of development cooperation as a technical game results in the very antithesis of "self-determination," as that term is employed in the officially recognized dispostif of emancipation. How is this possible when the entire world insists that the central objective of development cooperation is to establish independent self-reliance among those people who receive development aid?

The translation chain selected here is connected with many other chains in a worldwide network. Not all of the individual nodes in this network have the same meaning or significance, even if locally appropriate translations exist everywhere. Some nodes are able to define the procedural rules of the technical game in such a way that others are forced to follow them. Bruno Latour has introduced the term "centers of calculation" to describe such nodes.[30] These nodes are centers of calculation in the sense that they are concerned primarily with making the world calculable and thereby controllable from a distance. They are also centers of calculation in the sense that they are accountable to a general public, which in turn allows for their institutionalization, a process that requires a kind of calculation as well.

A center of calculation is thus an institution that collects *far-fetched facts*. This could be a natural history museum, which collects, classifies, analyzes, and documents all kinds of different species, or a cartographic center or a national weather service. However, centers of calculation also include the Colonial Office and the Royal Anthropological Institute in London, the World Bank and the International Monetary Fund in Washington, D.C., the Normesian Development Bank in Urbania, and in principle every node in a network that collects data about distant worlds (such as the anthropological Human Relations Area Files). The formal organization of every large technological system—a waterworks, a railway company, an airline—is also a center of calculation on a small scale. Centers of calculation produce new knowledge by bringing together things that in reality exist only in separate locations.

This in turn makes it possible to discover connections that would not have been uncovered at any of the individual locations. The knowledge produced by a center of calculation seeks to overcome its dependence on distant and resilient bodies of local knowledge by making them controllable. Its means of control are a variety of techniques that allow a center of calculation to appropriate the (always superior) local contextual knowledge of a social world without simultaneously being subjected to the restrictions of the locality.

It is possible to imagine the El Escorial Palace near Madrid, completed in 1584, as an archetypal center of calculation. Whereas Charles V still relied on control through personal presence and thus moved continuously throughout his enormous empire on which "the sun never set," Philip II did not travel at all. He held the empire together from the center, from his desk at El Escorial. For this purpose, he established an administration of civil servants, whose elaborate system of record-keeping allowed the center to monitor and direct the provinces. The centers of calculation that deal with development cooperation today are faced with a similar problem. The trickiest aspect of this problem is the possibility that each local context appears to comply with the guidelines prescribed by centers of calculation but in reality undermines them. Culture comes into play at this point as a dispositif of reflection and thus as a strategy of subversion. No formal means and no amount of power can eliminate this problem. This is why centers of calculation have a pressing need to outwit local dispositifs of reflection and to integrate them into their own programs. In this sense, they seek to dissect foreign cultures into "sociocultural factors" and to make predictions about the possibility of influencing them.[31]

My argument can be summarized as follows: The institutional rules of the field, the power constellations, and the irresolvable contradiction between the universalist and the relativist sides of the narrative of progress contribute to the perception of development work as a technical game. The hegemonic enforcement of the technical game by centers of calculation renders invisible the cultural and political aspects of those definitions of reality that enable the technical game to function at all. This serves to deny the dispositif of reflection of those cultures that are expected to assume responsibility for their own development. In my opinion, the notorious lack of sustainable development in sub-Saharan Africa can be traced to the fact that development cooperation is translated into a technical game.[32]

End of Report by Samuel A. Martonosi

SEARCHING

4 Interstitial Spaces

Preliminary Remarks : Edward B. Drotlevski

In his account, Johannes von Moltke (Normesian Development Bank, NDB) presented the aim of the project: Each of the three Ruritanian regional capitals has had its own central water supply since colonial times. Operations and maintenance of such a facility cost money, which must be covered by consumer payments. In Baridi, Mlimani, and Jamala, the share of the produced water that is paid for is so low that the waterworks could not survive financially. Obviously, something had to be done. Without contradicting this, Julius C. Shilling emphasized that the project set up to correct this problem was experiencing considerable difficulties and as a result could itself be facing failure. Shilling identified the main cause of the project's problems as what he called switching scripts. The official script (O script) ascribes autonomy and responsibility to the Ruritanian executing agents. Behind the scenes, however, the unofficial script (U script) applies, according to which the Ruritanian executing agents, as a result of their supposed incompetence, are more or less passive recipients of a service that is best determined by the financier and consultant. The inconsistencies that arise from switching between O and U scripts lead to all sorts of hypocrisy and coercion. Shilling was banking on the idea that the best way to stay afloat under such circumstances would be to have access to reliable facts and figures that promise immunity to the power-related distortions of script-switching.

Martonosi, on the other hand, believed that the greatest difficulty in development cooperation results from its reduction to a technical game with facts and figures. This game is based on the unquestioned basic premise that it is possible to comprehend reality objectively, that is, free of cultural influences. This self-evident and thus invisible assumption makes transcultural negotiations about the basic problems facing a project appear unnecessary and, according to Martonosi, proves to be a cleverly packaged form of hegemony. Of course, it became apparent in Martonosi's remarks that he consistently deconstructed the assumptions about reality posed by other players in the field, but at the same time attempted to construct an accurate image of the other players' constructions of reality, that is, an image that promised fidelity to reality in terms of correspondence theory. The essential difference of Martonosi's account from the first two reports therefore lay in its external referents and its implied message: "It is like this, but it could also be different."

I have collected and analyzed material from these three interviews at the descriptive level of "talking about." One of the most significant basic premises of anthropology since Malinowski—which became more authoritative through Wittgenstein's writings—is that the meaning of words and sentences can only be determined through their usage. This suggests that one should progress from "talking about" to "talking in order to." To this end it is necessary to go into the field, that is, to where the people are talking, in order to spur changes within the field in which they are talking.

On the Way into the Field

Baharini, Tuesday, September 2, 1997

Early in the morning on September 2, 1997, I met Shilling and Martonosi at the airport to catch the Tuesday plane to Baharini. The Airbus A 300 was named "Audrey Hepburn." My two companions turned without a second thought to enter the small business class cabin at the front of the plane, where they went right to their regular seats without looking left or right. The plane took off on time and shortly after takeoff most of the men in the half-empty cabin—not a woman to be seen—were hunched over their files and laptops. In contrast to what I am familiar with from flights to Johannesburg, Nairobi, and Accra, it seemed that on this flight only a few of the exorbitantly expensive seats of the business class were occupied by businesspeople who were paying their own way.

The sense of affiliation among business class customers went beyond a chance community of passengers. I was aware for the first time of seeing in the flesh the actors of the worldwide organizational field of development cooperation, from whom I have already learned a lot through interviews. Most of them were headed for a ministerial office in Baharini, where they would discuss projects. Some would venture to take the next step and "get into a project." The most significant medium for their activities is—as was obvious from one glance around the cabin—the report. Reports contain a multitude of lists, tables, calculations, graphics, organizational diagrams, and flow charts. The key art of this epistemic community evidently consists in being able to read and assess reports. To learn this art, it is necessary to be on site every now and again. The development travelers embark on their transcontinental flights in order to get a close-up look at the project activities and results. They try to confront the paper reality of the reports with the physical reality of the projects.

I talked with Martonosi during an uninspired lunch as we crossed over the Mediterranean. He said that the confrontation between paper and reality goes roughly like this: When they arrive at the project, the travelers collect, first and foremost, additional papers containing additional tables and lists. In passing they hear stories about how the project is progressing and they view a few technical facilities and are told how these are supposed to be used. Looking at a functioning well, however, does not say anything about the frequency of power outages, says little about how regularly they are serviced, and does not give a clue about changes in the groundwater level as a result of the facility. The freshly painted, small wooden shed to protect the well facility is supposed to demonstrate the discipline with which it is maintained, but it is impossible to see that discipline itself. And seeing a new computer in the accounting department does not reveal anything about the executing agencies' collection efficiency. Instead of the desired juxtaposition of representation and reality, all that is juxtaposed are further representations: More abstract, generalized, far-reaching representations are related to smaller, less abstract ones.

In the long term, representatives of development banks and consulting firms are concerned with building up first-hand practical knowledge that is useful in better reviewing the plausibility of reports from their "Here" (for example, in the Normesian Development Bank in Urbania) about a "There" (for example, municipal waterworks in Ruritania). It could also be said that the development experts receive reports full of lists, tables, and calculations that make no sense at all if the experts are not familiar with the corresponding pictures and narratives. Consequently, they occasionally set off on their way in search of these pictures and narratives. Rising in the hierarchy of a business or a development bureaucracy, of course, means less frequent trips into the field and, consequently, increasing dependence on existing documents and records.[1]

Later in the afternoon the captain announced that he would not be taking the usual route along the Nile valley over northern Sudan, but would instead fly farther westward over El Obeid (Al-Ubayyid) and from there over to Malakal before resuming a southern course over Ethiopia. For me that meant that we would be crossing the Nuba Mountains of South Kordofan, where I had lived for three years. There was not a cloud in the sky. I hurried into the cockpit—which was sometimes possible before September 11, 2001—so I could finally catch another glimpse of my lost paradise, if only from the air. The captain—a native Nigerian—found it extremely exotic that I could name the mountains and rivers below us, so he offered me the third seat in the cabin. He knew that southern Sudan was caught up in an armed struggle and he thought that this was also the reason why he had to fly over Ethiopian territory as far north as Malakal. But he had never heard anything about the genocide in South Kordofan. I saw Mount Lebu for the first time since December 1983, recognizing the mountain immediately even from this bird's-eye view. From letters I received from Khartoum, I knew who had been left behind on the high plateau of the mountain, living—at the moment I was flying overhead—in stone-age conditions and desperately trying to avoid the horrors of ethnocide. When we were over Malakal I had to return to my seat since a snack was being served in the cockpit. It was hard for me to go back to the development experts. Most of these experts consider the African wars to be a return to the barbarian past rather than innovative reactions to the presumptions of mismanaged modernization that they themselves helped to bring about with their projects.

Once we entered Ethiopian airspace I dedicated myself to the documents I had brought with me. James Wolfensohn, president of the World Bank, had announced the goal of increasing his bank's success ratio to 75–80 percent of all projects within only a few years. The 1996 success rate of 71 percent was already a significant rise compared with the 61 percent of the six preceding years. In the tenth Development Policy Report (report period 1992–1994), the Normesian Ministry for Development Cooperation (MDC) announced to the parliament that 76 percent of NDB projects were successful and that 13 percent had some positive impact. Interestingly enough, this cannot be compared offhand with the success rate of the Normesian AOD (Agency for Overseas Development), since they used different bases for calculation. Depending on the interpretation, either 52 or 78 percent of the projects were successful; 17 percent had some shortcomings, but still showed some positive findings.

Official discourse, however, also revealed some nuances. Two evaluation offices that are affiliated with the World Bank, yet retain their independence, came to the conclusion that for 50 percent of World Bank projects it was highly questionable whether they would yield any long-term benefits. Only

39 percent of the projects included a component for building up the institutional capacities of the loan recipient—there is significantly no talk of success here. As early as 1988 a similar study that had been commissioned by the US Agency for International Development (USAID) came to the following conclusion: The main cause of the great failure to "generate self-sustaining improvement in human capacity and well-being" lies in the fact that almost all individual projects are conceived in such a way that, despite the frequent rhetoric to the contrary, they are required at all costs to achieve short-term, measurable increases in productivity. The authors, in contrast, argued:

The design process should be based on documents which explain that the *highest priority for the project is to build capacity and not to achieve highly visible, short-term results.* Until donors begin to define capacity-building as the primary objective of the project, it is unreasonable to expect TA [technical assistance] personnel to interpret their role as one that extends beyond the performer model. Likewise, evaluation will continue to reinforce the performer and project approach that predominates in design documents.[2]

This is quite a remarkable statement considering that the water project I was planning to visit was evidently oriented around one single performance criterion: increasing the collection efficiency. In the same article I found an even more explicit statement:

Many donor-supported projects are based on the assumption that the efforts initiated will take on a momentum of their own, or at least that the host government will continue to support them. In fact, the development landscape is littered with the remains of projects that died when donor funding ended. Although these efforts were intended to foster a process of self-sustaining development, they provided little more than a temporary infusion of assets, personnel, and services.

This text was intended for use by development practitioners and policy advisors and had been based on research conducted by a private consulting firm in cooperation with a university on behalf of USAID. These observations can therefore hardly be dismissed as the prejudice of an ideologically biased critic. The study even found that large segments of the population in the forty-seven sub-Saharan countries were worse off in the late 1980s than they had been thirty years earlier. Finally, it stated that rather than an increase in the capacities of African institutions, the result has been a greater dependency on aid.

The tone gets harsher if you look at the relevant literature from outside the field. Here it has been assumed as virtually self-evident that almost all development projects in Africa remain ineffective at best; by no means do they satisfy their objectives. It is basically assumed that current conditions in Sudan and Ethiopia, for example, have arisen as a consequence of development cooperation:

The developmentalist transformation of the Third World has largely produced chaos and poverty.

The debt crisis, the African famine, and widespread poverty and malnutrition are only the tip of the iceberg of the development record.

The development fraternity has been casting around for several years for alternative approaches with mounting evidence of resources wasted in ill-conceived, frequently centrally imposed schemes that have not only failed to improve matters in lesser developed countries but have on occasion made them worse.[3]

It is interesting that even radically divergent appraisals agree as a rule on one point: To the extent that failures and errors are admitted at all, they assume that the cause is obvious. Most interventions and infusions are not accepted because they do not fit in the world in which they are intended to bring about something desirable. The blueprint approach assumes a prerequisite omniscient expert, who implants a universally valid model—a standardized package—and thus does not need to become personally involved, much less learn anything. The issue is quite different with the so-called process approach:

But the implementation of such an approach [the process approach] calls for an admission on the part of "experts" such as ourselves that we do not know everything and, furthermore, that we are prepared to learn from our mistakes. But most importantly, this model asserts that development involves personal transformations that can take place only if individuals themselves are intimately part of the process—that is, if they shape it and are transformed by it.[4]

That was not going to happen in the case of the Ruritania waterworks, however. Indeed it seems baffling why this costly experience has not yet resulted in any notable changes in practice—despite the obvious agreement that it is impossible to transfer preexistent models.

Late in the evening the plane landed right on time in Baharini, and Shilling gave me a preliminary briefing: Baharini is neither an industrial or trading center nor an administrative or cultural capital. With a population of roughly two million, it is a relatively small, sleepy city. Here you are constantly reminded that you are in a place that is integrated not into the world market, but into the system of planned and sponsored development. For most of the business class passengers, they could have landed in any African capital. From their perspective these are randomly interchangeable hubs in the global network of development cooperation. On the scale of possible tests of patience in crossing border controls, Baharini is a relatively harmless place for travelers with "good" passports. The numerous luggage carts, taxi drivers, and other tourist guides gathered at the exit are much less pushy than in Cairo or Lagos.

Deregulation

While we were waiting at the carousel for our luggage to come out, and were slowly but surely breaking into a sweat because the air conditioning in the arrivals hall had evidently broken down, the power suddenly went out altogether. It took quite a while for the airport personnel to assemble a sufficient number of petroleum lamps. During these long, dark minutes our airplane shone as the sole source of light at the entire airport of Baharini. Even the control tower and the runways did not appear to be connected to any emergency power generator systems. As soon as the petroleum lamps were installed, the lights went back on. As Shilling explained, airport technicians had managed to switch from RURESCO, the parastatal power company, to their own diesel generator.

September is dry season in Ruritania. As in past years, most of the river-beds had dried out, causing the reservoir of the hydroelectric power plant to sink to a dangerously low level. Since the technological infrastructure of the metropolis is highly sensitive to power outages, there is an emergency system that can temporarily supply the city with power through diesel generators. A ship has been lying at anchor in the port of Baharini since August, laden with the needed fuel ordered by RURESCO. In order to import this fuel, RURESCO has to pay a sizable sum in import duties. Corresponding to the primary principle of the structural adjustment—"to get the prices right"—the state is not supposed to subsidize the price of electricity through remission of customs duties. The power supplier, however, is unable to pay the duties. The reason for this is widely known and can often be read in the newspaper.

The Ruritanian government has been urged by the World Bank Group to use political regulatory mechanisms to make sure that RURESCO, as the sole power supplier in the country, covers its operating costs according to principles of private enterprise. This means, for one thing, that the customers must pay an adequate price. With respect to private customers, at least, it seems to be generally possible to assert this business policy by interrupting the supply if necessary. With respect to state authorities, however, RURESCO is evidently powerless. Debts in the "institutional customers" category have meanwhile reached a level that makes commercial operation of the company seem questionable. The sinking water level in the reservoir and the existential dependence of the city on electricity have strengthened RURESCO's negotiating leverage considerably. (I later heard that an agreement had been reached on a solution involving customs duty remission for political reasons.)

As the luggage conveyor belt was starting up and the first traveler heaved his black hard shell suitcase onto his wobbly luggage cart, Shilling jested, "I'll bet that's the man from the World Bank who hurried by to make sure, at the last minute, that the fuel is not imported duty-free." (In the city we later heard: "They want us in Africa to vegetate without electricity or water.") In fact, however, virtually all development experts are more or less involved in spreading privatization and deregulation as the last-remaining idea that appears to be worthy of global dissemination. Within the structural adjustment (as von Moltke explained to me), the point is to reduce to an absolute minimum the areas of responsibility the state assumes as guardian of the common good (devolution) and to increase the distribution rationality of the market to the absolute, barely justifiable maximum (privatization and deregulation). To this end even the organizations that define the institutional scope of the new order are supposed to be subjected to the laws of the market as far as possible (privatization and commercialization), and the democratic decision-making structures are to be organized so that the problems are solved and paid for where they arise (subsidiarity, decentralization, deconcentration, participation, and empowerment).

The World Bank, the International Monetary Fund, and also the various national agencies for development cooperation of the entire Western world currently claim that they would offer all the funds at their disposal to create not only the necessary economic policy conditions (such as currency and tax policies, subsidy and distribution issues, importation licenses and quotas, etc.), but also the necessary institutional prerequisites for the desired structural adjustment. Our water project fits into this context. In concrete terms, this means that the water supply—like the electricity supply—must be structured such that it pays off. If this condition is not met, then the organization must be modified accordingly so that it works without any subsidizing transfer payments that would come from taxing other commercial activities. The magic formula is the cost-benefit analysis.

As routine experts who were "going to their projects," we did not have to worry about finding a taxi. The project chauffeur greeted us with unobtrusive friendliness and led us through the crowd to a silver Toyota RAV with air conditioning and a stereo, and which still had an intense "new car" smell. In contrast to the city streets, the dust-free four-wheel-drive vehicle was so surprisingly quiet and pristine that we felt alien. I wanted to smell the musty air of the tropical capital city, but the driver didn't let me open the electric windows. Without any further questions he brought us to Hotel Tulip, where my companions often stay since their former favorite, Hotel Africa, has become rundown. The Tulip has the reputation of being respectable because the British woman who runs it has prohibited prostitutes from entering the hotel bar (an otherwise common occurrence in Baharini).

Translation Work

Baharini, Wednesday, September 3, 1997

The first day of work began with the familiar attempt to set up appointments by phone. According to a 1997 calculation by the telephone company, 45 percent of attempted telephone calls in the landline network in Baharini result in a successful connection. But meanwhile there is also a wireless network that operates far more reliably because—as Martonosi explained—the technical system of this network has reduced its dependence on human resources to a minimum. We reached our first contact by cell phone and walked over to meet with him.

Mr. J. is an engineer and the man at the water ministry in charge of all Normesian projects. Today's meeting was about problems regarding customs clearance for water meters that had been imported from Normland and were now waiting in port. Martonosi said that we were very fortunate since J., whom he has known for five years, is difficult to reach—like all his colleagues. In 1992, during a joint trip around the country, the two got to know each other somewhat better, and J. openly explained his frequent absence from his workplace. His income is so low that he cannot support his family. While his wife breeds chickens in the small yard behind their house to sell them to cookshops, he procures the feed and besides that does various sideline jobs. That is why he is out of the office every day from 11

A.M. to I P.M. Next to his official work telephone, which appeared to have been on his desk since British times, was his cell phone. Thanks to this device, he was able pursue his diversified economic strategy largely from his desk. This was particularly helpful in view of the constantly threatening gridlock on the city streets.

During the conversation about duty exemption for the water meters for Baridi, Mlimani, and Jamala, there was also talk about future water projects in another four cities and with different funding sources. The ministry had been called on by a multilateral financial backer, who remained anonymous, to review and possibly expand a list with possible applicants to carry out this new project. It was never stated explicitly, but in the course of the conversation Mr. J. got the idea that S&P was interested in this new project and that he could help. In this context he handed Martonosi a business card and said that he himself worked for a private Ruritanian engineering consulting firm, which evidently wanted to participate in development cooperation. However, a Ruritanian consulting firm can participate in a development cooperation project only by cooperating with a Western company, as Western investors tend to trust companies from their own country or cultural sphere. From the perspective of Baharini's professional elite, this practice of allocation must seem like the height of postcolonial arrogance.

Martonosi explained to me on the way home that several Ruritanian experts can be hired for the same cost as one person-month of a Western company. He said this is common knowledge here. Especially people like J. have an eagle eye in this regard, since the information collects on their desks. In contrast to J., the directors of the waterworks in Baridi, Mlimani, and Jamala might not have an overview of the total amounts, but from their perspective as executing agents the matter does not look any less scandalous for another reason: The financier puts a certain sum at the executing agents' disposal. They in turn have to use this to pay the fees for the Normesian consultant, which seem obscenely high within the local framework. (For example, the initial offer from S&P for 2.5 million dollars included 1.8 million for fees, travel, and accommodations for the Normesian experts.) This inevitably leads to considerable tension in the cooperative effort. The options for dialogue are seriously restricted if one of the interlocutors earns 600 dollars during the joint workday, whereas the other would have to work an entire month or longer to earn the same sum. This massive income disparity impairs the sense of reciprocity that is necessary for productive communication, which is reason enough to question any understanding that is reached under such circumstances.

Chapter 4

According to Martonosi, the water minister announced in a public speech during the so-called Water Week in the city of Jamala, in the fall of 1996, that a large share of the country's development problems were tied to the fact that the financial backers give their contracts to Western consulting firms. These companies had neither the necessary expertise nor a heart for the country, he said, and consequently they had primarily their own profits in mind when designing the projects. The minister thought that was the real reason why so many development projects had no lasting positive results. Ruritania would not head down the proper development path until it took matters in this area into its own hands. This interpretation by the minister was also remarkable because it agreed with the public statements of Western development agencies. In Urbania both the ministry and the development bank have said that the most important thing development cooperation can do would be to finally put the responsibility into the hands of the stakeholders who want to improve their own situations.

The centers of calculation of the donor countries basically feel that the reason why development measures have not been adopted—at least not as expected—is essentially the attitude of the African elites, who often evade their responsibilities. This cannot be stated publicly, however, which is why one either diplomatically chalks it up as one's own mistake or denounces it as an inexplicable failure in the past. On the surface it then appears as if everyone in Baharini, Washington, and Urbania agrees. If projects do in fact go awry—according to the obvious conclusion from the perspective of people like J.—it had to be due to the Western consultants.

Shilling and Martonosi were able to arrange a late-morning appointment with the undersecretary of the water ministry. They wanted to find out how far the legal prerequisites of their project had developed in the meantime and the current status of the implementation provision. Because they wanted to introduce an incentive pay plan and hire new staff as soon as possible, they were interested specifically in the corresponding "special authorization," which the undersecretary had promised in October 1996 in Mlimani. I had to wait for the two of them in a café during their entire meeting since, I was told, my presence as a neutral observer would have hindered the negotiations. Negotiations require compromises and, with that, a certain loss of face for one party or the other. It can be expected that participants will show any willingness to give in only as long as everyone present also has something he or she could potentially lose. As soon as someone is there who has virtually nothing to lose by participating, it destroys the necessary reciprocity of the group. A neutral observer is also a dangerous witness, whereas an engaged participant will not say anything that contradicts the group's common interests—at least as long as the group continues to have a common interest.

About two hours later the two friends walked into Café Zebra on Independence Avenue, where we had a snack together. Shilling was sarcastic from the start. They had waited an hour and a half for a fifteen-minute conversation. They had paced up and down a windowless corridor, where he had suffered from the sauna-like temperatures, and Martonosi, from the claustrophobic size of the space. Meanwhile they had memorized the signs on the doors, which are usually out-of-date in Ruritanian office buildings anyway. The undersecretary, an elderly professor of medicine with one of those 70° F offices, evidently did not spend much time on details. Most important, however, he was painfully aware of the fact that a Normesian undersecretary would hardly ever debate questions involving issues of sovereignty with a foreign consulting firm that has been contracted to reform the pay plan of the waterworks of Mercatoria. He made it clear to the two development experts that things would take their proper course, that they would do better to keep to their "own affairs," and that the Normesian people had a poor reputation for being overly inflexible.

Following the Mlimani workshop in October 1996, it would have been essential to make the pilot project possible through special authorization, but that was not forthcoming. This morning, the attempt to directly address the highest authority about this went awry. The two consultants drew the same conclusion from this disillusioning morning: They should exert even more pressure than before on the executing agencies to offer some solutions to the bogged-down situation. S&P apparently had not received any feedback on the midway review of the project, which I had heard about repeatedly.

The service that contractors sell in this context thus is apparently the mediation of social worlds in the global arena of development cooperation. To this end they usually physically transport themselves from one social world into the other and later return. To this extent, they are specialists of spatiotemporal transitions in interstitial spaces. Members of the social worlds between which contractors mediate are themselves not able to communicate with each other directly, since they do not, in a figurative sense, share a common language. They are also not permitted to negotiate on certain subjects since this would damage an existing order and role allocation that on another level needs to be protected.

Imagine if this morning, for example, the undersecretary had met with an appropriate counterpart instead of with Shilling; ideally that would be his Normesian colleague from the MDC in Urbania. In that case the issue of status would be resolved, but neither of the two would know precise details about the actual project. Nor would they be able to talk about project details with the help of expert prompters whispering in the background, since they might embarrass each other in the process. This would threaten the successful outcome of their other negotiations, which must be more important to them. Their talks could sensibly take place only at a higher level, at which the general framework and not individual projects are clarified.

The other obvious solution, that the two consultants should have gone not to the undersecretary this morning, but to administrative officials in the ministry at an accessible level in the hierarchy, would also not have worked. Competence, power, and decision-making authority are distributed in Ruritanian bureaucracies such that those who are formally responsible have neither the expertise nor the power to make decisions. (For this reason it was possible to talk to Mr. J. today only about customs matters, which is why Shilling did not even bother to come.) At some higher levels the necessary know-how exists, but here as well, decision-making powers are missing. And at the level where decisions can finally be made, the necessary expertise is lacking. In the present case it was also true that most expert key figures in the ministries—Mr. N. and Mr. S., who are also on the boards of the waterworks in Baridi and Jamala—felt skeptical about the project because it undermined the role of the ministry and thus their own significance. These two elderly gentlemen had climbed in the hierarchy of the ministry during the socialist era. Today they made no bones about their conviction that the whole deregulation and decentralization campaign was misguided.

My two consultants thus did not find any alliance partners in the water ministry who had any interest—out of conviction, friendship, or for career reasons—in giving the two of them confidential information or in assisting them in other ways. Martonosi told me that for a while in the fall of 1996 it looked as if he had found such a partner. At that time he had met an astute ministerial staff member at the third level of the hierarchy whose career had been blocked by his direct superior, one of the two gray eminences there. For this reason, he was interested in attracting the attention of the top level of the ministry, especially since he knew that the minister was more open to reform than his evidently somewhat blockheaded boss. Nothing came of this alliance, however, since this man changed ministries.

Another reason why the impossible feat has thus far remained impossible is that the necessary degree of consensus among the parties involved is lacking. Since this is a case of cooperation under conditions of heterogeneity, "consensus" must be defined more precisely. It is not a matter of establishing a community that agrees on all aspects of interpreting the world. On the contrary, in order to accomplish anything under these conditions, as many opinions as possible must remain factored out and only those that are absolutely necessary in pragmatic terms should be discussed. In this way a *trading zone* is created, which reduces the conflict potential to the unavoidable minimum and assures that the process can continue. The common technique for remaining as independent as possible of the shaky consensus of the players involves formalizing and standardizing procedures.

As the expert of the interstice, the development contractor manifests his mastery of this art by defining the situation and problems in a way that seems acceptable to all sides, that is, sufficiently open for local reinterpretation but at the same time reliable enough for a concrete project to be predictable, translocally attributable, and independent of culture- and interest-related fluctuations in interpretation. As a mediator, the contractor must develop or identify standardized procedural models and artifacts—"standardized packages" and "boundary objects"—that are solid enough to circulate back and forth between social worlds undamaged as "immutable mobiles," but also flexible enough to be adaptable to various local contexts. Under the circumstances that apply for our project, a technical game was agreed on which took these prerequisites into account, as Martonosi explained, but evidently things were not going well. At least one reason for this was that the local reinterpretations of the standardized packages and the boundary objects turned out to have been too extensive.[5]

After my two exasperated companions finished their espressos, we pushed our way through the bustling throng along Independence Avenue toward the hotel, passing the heroes monument, where everyone tries to sell anything and everything and where pedestrians instinctively pay close attention to their bags.

The Technical Game as a Code of Reciprocity

The evening of the first day of work we met the managing director (MD) of the waterworks of Baharini in an Indian restaurant on Independence Avenue, where a decent dinner for four costs more than a waiter's monthly wages. My two companions were anxious to hear what the MD had to say about the latest developments in the water sector, but they also ran into an old acquaintance, with whom Martonosi had worked for two weeks back in 1992. The three specialists filled me in on the background so I could follow the conversation.

The devastating upshot of the socialist water policies and the centralist administrative apparatus was already apparent in the 1970s and can be summarized in one sentence: "Now the water is free of charge, but it has stopped flowing." Consequently, an initial rescue was ventured in 1984. The National Urban Water Authority (NUWA) was founded as an autonomous, commercially operating, state-owned agency whose jurisdiction initially pertained only to Baharini, but was supposed to be expanded little by little to all other Ruritanian cities. This experiment had two inconsistencies from the outset. First, when urban councils were reintroduced, an amendment that same year put them in charge of the municipal water supply. This stood in contradiction with NUWA legislation that had been passed at the same time. Second, the founding of the NUWA was not accompanied by a consistent dismantling of the old structures. According to Martonosi, this is a standard pattern in the Ruritanian state administration, which he calls "organizational extension": a new entity is set up to replace an older, dysfunctional one, but then ends up as an extension of it. The NUWA never managed to become an independent, much less national, agency; instead, within the scope of reorganization in 1997 it was restricted to the Greater Baharini area and renamed Baharini Urban Water and Sewerage Authority.

Between 1984 and 1997 the minister was supposed to appoint the supervisory board, which enabled him to prevent the complete separation of the ministry and NUWA. Even more important was the fact that until the early 1990s the minister for water, energy, and minerals was responsible for both water rates and electricity rates. While on the one hand he focused on keeping drinking water affordable, on the other hand he also made sure the rates for electricity rose parallel with production cost increases. Behind this unequal treatment of water and electricity lay the different symbolic force of the two commodities. Water is associated with the basic needs of the poor, the sick, and children, and considered an absolutely vital prerequisite for survival, much like air is not supposed to be subject to the laws of the free market. Electricity, on the other hand, is associated with surplus, luxury, and progress—in other words, a good that must be paid for. If a ministry is responsible for both of these public goods, it is possible for the "ethical" commodity to be subsidized through the less ethical.

But because the costs of supplying water in Baharini are largely electrical costs, the minister systematically maneuvered the NUWA into an escalating debt spiral. At the same time, he neglected to introduce a transparent system of subsidies. Thanks to this self-created cause for the NUWA's debts, the minister, as chief crisis manager, had a lot to worry about. However, it was also precisely the nonfunctioning structure and lack of transparency that gave him a disproportionate amount of power and influence. The NUWA was at his mercy, so to speak, since he was the only one who could remit its electricity debt. Under these circumstances the waterworks were by no means financially autonomous, which had been the official aim when the NUWA was founded in 1984. The MD laughed bitterly, saying that the city's contractors would no longer hand over even a bag of cement if the buyers from NUWA did not pay immediately in cash.

During dinner, the hard-nosed MD received news on his cell phone that he had to give a TV interview the following day. He was supposed to talk about the possible connection between the numerous cases of cholera that had broken out in a certain district of Baharini and the non-functioning water supply there. It had been hit hard recently, not only by drought-related power outages that continue to paralyze the water treatment plants. Even more serious and direct are the effects of the low water level of the river, the source of almost all the city's drinking water. Water rationing, which had always been necessary, had now reached a level that would have led elsewhere to a state of emergency. (I was told that Baharini had a daily water consumption of 90 million gallons, but the production capacity of the waterworks is only 60 million gallons. The discrepancy is actually much higher, because production can never reach full capacity because of technological deficiencies and because 30 percent of the produced water volume is lost in the system of pipes through leaks and other defects.)

The MD attributed this situation mainly to the notorious lack of funds and thus also to the lack of new technologies. (In the coming evening I was to witness the "lack of funds" rhetoric that Shilling had told me about during our conversation in Mercatoria.) Shilling was trying carefully to steer the discussion toward the issue of whether the necessary revenue to operate the waterworks might not best be generated by selling the water. But the MD quickly brought the discussion back to his own argument, namely, that bringing in money from the customers first requires money for computers, training courses, improved customer data, current and if possible digitized maps of the pipe system, water meters, sonographic detection equipment for leaks, and so on. No matter how you look at it, the MD saw it as a matter of a "technical fix."

In order to better appraise the chances of their own projects, my two companions asked the MD over their last glass of South African wine how they should interpret a call for bids in the newspapers: Whereas they had heard from the undersecretary that morning that commercialization of the municipal waterworks would be possible in the near future—though no details could yet be mentioned—the papers reported that a private operator for the waterworks in Baharini was being sought. The water ministry was said to be considering an operator that would assume responsibility for Baharini according to the BOOT principle (build, own, operate, transfer). As a party directly affected, the MD confirmed this news with a smile. He doubted that this strategy would lead to a business deal unless the charges that the company billed its customers for distributed water were reimbursed by the city government. But that would lead only to the city being driven to bankruptcy instead of the waterworks, since the city could also never succeed in collecting payment for the water bills.[6]

While Shilling and Martonosi were drinking another whiskey at the hotel bar, I recorded my deliberations in my field diary. Martonosi had tried to explain to me in Urbania that development cooperation is reduced to a technical game, in which all that matters is the distinction between effectiveness and ineffectiveness. In his view, the main reason for this reduction lies in the existing power disparities and hegemonic claims of those who play the role of the so-called donors. Yet that evening it was the MD of the Baharini waterworks who had insisted on the technical game as the only one in town, and out of courtesy Shilling had not contradicted—which suggested to me that there might be other reasons for the dominance of the technical game.

One of the greatest difficulties facing the Baharini waterworks seemed again to be precisely what Shilling briefly described for the waterworks of Baridi, Mlimani, and Jamala. The problem is a mechanism that leads existing lists of facts and figures to become invalid during the period of their use; in other words they consume each other and themselves from the inside. This is why Martonosi speaks of a mysterious *list autophagy*. The supposedly trivial ability to locate one's own customers in order to provide customer service and collect money for those services rendered is evidently insufficiently developed. While one might think it is important to get at the root of the problem and first find out what is causing this strange list autophagy, energy instead seems to be expended in avoiding this. The solution triad of the technical game—loans, technology, expertise—comes up again and again with unquestioned certainty. I now believe that this suspicious factoring out of the question of causes might be due to something other than merely the hegemonic dominance of the technical game, as Martonosi seems to think.

What if we were to imply that list autophagy were caused by rational calculation? That is, suppose that the workforce of the waterworks were deliberately falsifying and misplacing the lists because it brought them some financial advantage. For example, someone could promise a customer who is not entered on the list of registered customers that she would be kept off the list in exchange for a portion of the money that she would have to pay if she were on the list. This argument would imply that here we are dealing with a corrupt group of people, and the MD, as the main person responsible, would automatically be discredited. This would not be a good basis for sitting together at a table to discuss the obvious problems.

Or what if we assumed that list autophagy were a problem of culture? In the evolutionist variant of this explanation the MD would have stood there this evening as the representative of a culture that cannot yet cope with the principle of documentation. Although it has indeed adopted the use of writing, it has not yet created a *culture* of writing. This would result in an impossible situation, one that had to be avoided at all costs. The relativist variant of the culturalist explanation would also quickly put us in a paralyzing state of indecision. Claims that definitions of reality are formed largely by frames of reference would build an insurmountable wall between people about to agree on a common definition of reality valid beyond any boundaries that might exist. In that light, the MD would have stood there this evening as the representative of a culture that because of its different definitions of reality brings forth very different forms of exchange between the parties active in the drinking water sector. The two people with whom he was talking, in turn, could hardly consider these forms to be rational, given their culture-specific frame of reference. This again would not be a good basis for sitting together at a table to discuss problems.

The possibility for talks to take place at all in heterogeneous trading zones is dependent on whether propositions can be made that are independent of any specific frame of reference. The only propositions that can satisfy this criterion are those whose truth-value is supposedly based solely on their correspondence with external reality. Ruling out the first two explanatory models of list autophagy—calculation and culture—leaves a third model as the only possible alternative: The waterworks have lost access to their customers because they lack the necessary money, computers, and other contemporary technologies and trained personnel. Only at this level can the game function. In the trading zones of development cooperation, experts of all nationalities and continents communicate on the basis of a universal rationality, factoring out all issues that would make it difficult to continue. They all have agreed on a bare-bones definition of the situation and the corresponding problems based on a kind of unspoken yet highly effective standard operating procedure. Thus the problem becomes a matter of eliminating any deficiencies that can be eliminated: insufficient credit, outdated and broken technology, lack of technological expertise. The function of the so-called technical fix, which Martonosi had described to me as a hegemonic strategy, was perhaps instead its indispensability as a *code of reciprocity* for the trading zone.

The story of the bid by the Baharini waterworks raises another far-reaching question that is gradually regaining topicality: What are the actual limits of privatization, which has been celebrated so euphorically as a panacea in the field of development cooperation? State bureaucracies fail to fulfill their responsibility of acting in the name of the common good, in particular as their staffs and strategic groups prevent them from doing so—this is the accusation vehemently leveled at African bureaucracies. From this perspective, the problem facing African countries lies in the fact that the visible hand of the state is being disturbed by the many invisible hands of the market. If that is so, then it seems somehow misguided or at least counterintuitive to praise privatization as the remedy for the public hand. Both variants of deviating from the correct and proper path—kleptocracy through too much market in the state, and market failure through too much state in the market—tell only half the story. If the central problem of social development lies in the fact that particular interests are inadequately combined into general interests, and if we presume in addition that this is a result of unrestrained utilitarianism, then it seems astounding to offer utilitarianism as the solution.[7]

Objective Data

Baharini, Thursday, September 4, 1997
On the morning of the second day of work, I was permitted to accompany Martonosi to the Prime Minister's Office (PMO). This is where in 1992 he had been told—as I already knew—of the possibility of operating the three waterworks on the basis of the Revolving Fund Act. And this is where the

measure was pushed through in 1994, more in opposition to the water ministry than with its support. This in turn was the situation that led the NDB to start financing the Organizational Improvement Program in 1996. Today Martonosi wanted to find out how, from the perspective of the PMO, the Ruritanian waterworks would be further decentralized and commercialized. During the taxi ride he told me that the person he had spoken to in 1992 was now the undersecretary in the Ministry of Defense; so instead he had arranged today's meeting from Urbania by email with a Danish sociologist who worked as a government advisor in the Prime Minister's Office. The Danish government advisor was very familiar with the details of the debacle surrounding the autonomy of the municipal waterworks, and he brought Martonosi to Mr. K., who, it was thought, could perhaps help further.

The overarching program to trim down Ruritanian's civil service, which was a precondition of the World Bank for certain loans, happens to be located in the PMO. The reorganization of the municipal waterworks was, as Shilling had already explained to me in connection with the workshop, only a small and rather insignificant aspect of this overarching program, which also included, for example, the huge public health sector and the even more expansive educational system. Mr. K.'s success is currently measured by the number of civil servants he is able to remove from the public payroll. It turned out that although K. has access to the personnel lists of all the municipal waterworks, he had not been informed that the three waterworks involved in the S&P project could already cover all their personnel costs themselves, as budget analyses have shown. Martonosi explained this to him somewhat hesistatingly, whereupon Mr. K. spontaneously decided to adopt the issue as his own. He was prepared to make sure the workforce of the three municipal waterworks would soon be removed from civil service payrolls. While we were still sitting at his table, he called various colleagues in the building and in the water ministry to convene an initial meeting the very next day to discuss the matter. He also reached N., the person in the water ministry responsible for all municipal waterworks in the country. The day before, Martonosi and Shilling had not sought to contact to him because he had been openly opposed from the outset to the concept of a "pilot project with special authorization."

As I was sitting there in amazement, Martonosi turned pale. On the ride back in the taxi, I asked him if his complexion change had anything to do with the fact that Mr. N. from the water ministry might now find out about his meeting in the Prime Minister's Office and possibly intervene. Martonosi said no, that although there was some danger of this, it was relatively improbable, since the two offices have up to now never acted in concert. Instead it was the surprisingly committed and quick reaction of the ambitious Mr. K. that had raised a probing question for Martonosi: What if the waterworks do have problems paying their personnel costs after all? Martonosi said that to some extent he had now assumed responsibility for the existential repercussions this would have on personnel and the municipal water supply, and that this responsibility was based on so-called facts and figures, of all things! No one who had any idea at all how these figures came about could really trust them. Aside from the unreliability of the elementary data, Martonosi explained, they are based purely on speculative projection. And although the planned expenditures could still be predicted with some degree of certainty (always excluding catastrophes), for calculating revenues they would have to rely on prognoses about increasing the collection efficiency. I added consolingly that in such decision-making situations one obviously required firm ground under one's feet, the construction of which cannot be questioned, at least not at the moment of making the decision. To be capable of action one had to suspend doubt and self-reflection. Skepticism could only be cultivated where it was appropriate, and then it should quickly be left behind again. But my academic commentary only made Martonosi even more nervous.

The validity of the following proposition: "The waterworks of Baridi, Mlimani, and Jamala can pay their personnel costs out of their income revenue" is presumably so precarious because it is not embedded in a network of other propositions that support and reinforce each other. Such validity can be determined solely on the basis of the correspondence between proposition and reality. Yet that seems to overload the concept of validity. Behind the apparently culture-free budget figures lurks a prognosis based on a culture-specific assumption: It is presumed that collection efficiency can be increased by having the waterworks pay their salary and wage costs out of their own revenue and by introducing an incentive pay plan. Behind the figures that have to be used to make such a decision lies an entire worldview, the validity of which is questionable for the local context. Precisely because of this uncertainty, both sides had made every effort to exclude it from the project negotiations. Now it was suddenly being made the foundation for a far-reaching decision. Martonosi felt faint when he realized how the overloaded concept of validity and the mathematical rhetoric of business prognostics had been shifted onto his shoulders, just because he'd run into Mr. K., who unexpectedly took the problem seriously.

Decentralization

While Martonosi and I were in the Prime Minister's Office, Shilling met with representatives of the European Union at the Delegation of the Commission in Ruritania. He wanted to hear their appraisal of the future legal framework for the urban water supply and find out how far the planned involvement of the EU in the water sector had come. During a coffee break on our hotel terrace we shared our experiences from the first part of the morning. EU involvement was to first manifest itself in a project aimed to improve the organization of the water supply in four Ruritanian cities, based on our model project. I realized that this obviously concerned the same project that Mr. J. had been talking about yesterday at the water ministry, when we were there regarding customs clearance. J. had mentioned a list of applicants and expressed his outrage that Ruritanian consulting companies were not being considered and that the so-called short list of the most promising applicants had not been drawn up by the ministry. People in the EU office in Baharini evidently also reacted to the list with indignation. The local EU representatives were of the opinion that because of their superior expertise they were the ones who should have drawn up the short list. In Brussels, on the other hand, officials thought the main thing was to distribute the project bids fairly to consulting companies from all the countries of the EU, so of course they felt the short list should be drawn up in Brussels.

After our coffee break I accompanied Shilling to the World Bank office. It was nearby, in Baharini's only skyscraper, where numerous international companies had their offices. It was not easy to get an appointment, since the staff of the World Bank very cleverly hid behind technical security procedures, rigid formalities, and professional receptionists, who were hard to get past if the person you were looking for did not have any interest in meeting with you. Shilling had been trying for over a year to find out precisely what the World Bank planned to do with the so-called Urban Infrastructure Rehabilitation Program, how far they had progressed, what repercussions it could have on work within the S&P project, and on what points they should coordinate their efforts.[8]

A young American who ended up sitting next to us—obviously fresh out of college—started by explaining his area of responsibility. He was in charge of public relations work for the World Bank office in Baharini and the accounting for all Ruritanian projects. He referred to his subject area not as programs and projects, but as the "country portfolio." Furthermore, he said, the program we were interested in of course had a local coordinator—as was true of all World Bank projects—who most certainly was the best informed about the program. Shilling dismissed the tip like a bad joke: Almost a year ago he already met with the Ruritanian coordinator, who proved to know next to nothing about the program. The young American responded sheepishly and for a moment didn't know whose image he should defend: his own, that of the Ruritanian colleague, or that of the World Bank itself. He settled for the World Bank and said that they were currently seeking someone new for the position of local program coordinator since they were in fact dissatisfied with the incumbent. He said the man would probably be replaced by Mr. S. from the water ministry, "a very capable man." We were then told that the best informed person in this matter was the present program director of the World Bank in Washington, D.C. It sounded as if the young banker wanted to say: "Not by any stretch of the imagination can I refer you to anyone any higher in the hierarchy." Shilling dismissed the comment with the same laugh, remarking that that man was in Baharini only two or three times a year, only for a couple days at a time, and he did not respond to emails on principle, which was why we were here today in the first place. The young man now showed visible signs of insecurity, and he finally started perusing his shelves for reports on the program. After Shilling left his passport as "security" he was allowed to borrow and photocopy a document from 1996 that was most likely to contain significant data.

We concluded the morning by taking a walk along the beach promenade to the port and from there we climbed up to our hotel. Shilling said that by now I could begin to guess what absurd problems one faces when trying to coordinate the countless projects of a developing country. Even our relatively small project on reorganizing the waterworks of Baridi, Mlimani, and Jamala was part of a complex network of other projects and programs that in principle should all be interconnected. Among them, for example, is the huge Civil Service Reform Program, which involves the reorganization of the regional and city governments as well as a drastic reduction in the personnel of the public administration. Within this framework there are also plans to reorganize all the ministries, the implementation of which has been ongoing since 1996 with varying degrees of commitment. One goal of this new order is to transfer competence for the operative tasks to lower-level authorities, which are supposed to be granted differing degrees of autonomy. That was the reason for the meeting in the Prime Minister's Office this morning. The three waterworks projects of S&P are thus affected by just this one framework in a number of ways.

There is also another comprehensive World Bank project, the Urban Infrastructure Rehabilitation Program, which we had just touched upon. This program targets fundamental questions of urban management and allows for smaller measures to be implemented directly in the cities of Baridi, Mlimani, and Jamala. Most of the ministries in Baharini also have so-called government advisors, whose job is to promote sector-specific reorganization. In the ministry of water, this job was to be done by an employee of the Normesian Agency for Overseas Development (AOD), whereas in the Prime Minister's Office, it was the responsibility of the sociologist from Denmark whom we'd met briefly this morning.

Shilling said it was a common error to attribute the obvious chaos in coordination to the lack or failure of centralized planning;[9] instead, he thought the problem was actually caused mainly by an "upward spiral," which we had talked about yesterday after the meeting with the undersecretary. In most other state administrations, things work just as they do between the levels of hierarchy in the water ministry. In a quasi-reversal of the subsidiarity principle, tasks and responsibilities are drawn upward until the decision-making authority ends up at higher administrative levels, levels that do not have the necessary specific knowledge, are not affected by the pressure of the problems at issue, and hardly ever have to directly endure the repercussions of bad decisions. The regional governments of Baridi, Mlimani, and Jamala, for instance, repeatedly ask what the administrative reform will entail in detail and precisely how it will affect the water sector, but no answer is forthcoming. They are regularly referred to one ministry or another in Baharini, but when they show up in the ministries with these questions, they are sent back to the regional administrative level. This ultimately makes the coordination of complex and wide-ranging processes prone to failure.

I interjected that this story sounded like a trick to me, which the Ruritanian elite uses to get rid of unpopular questions. People pretend they lack the competence or responsibility in order to assure themselves a certain degree of leeway for their actions. This defensive strategy must be a reaction to an extensive infiltration of the country by foreign-run programs that hollow out the authority of the existing administration in the name of the people's empowerment. Shilling was able to explain this theory more precisely. The confusion regarding who is responsible for what climaxes as soon as the donor organizations enter the scene. We had just gotten a small taste of this climax in the World Bank office: The major donor organizations are even less coordinated with each other than the Ruritanian ministries. In the case of this project, for instance, it should have been possible to find out about the status of the Civil Service Reform Program and the parallel establishment of commercially operating enterprises prior to the commencement of the project, in order to set one's own objectives in relation to that. To a certain degree the donor organizations were justified in saying that it cannot be their job to work out the coordination of various programs all the way to the implementation level. Not only would that mean an immense and costly outlay, but it would amount to patronizing the Ruritanian partner—and would in fact lead to a cynical extension of the so-called upward spiral.

But since the donor organizations nevertheless do consistently experience the lack of adequate coordination in Ruritania, they ultimately pass the buck to the consultants. This avoids the politically untenable state of affairs of having Ruritania run by the office of the World Bank. The job of coordination is shifted onto the player with the greatest and most direct economic interest in it. The consultants are the only participants who have an existential need to succeed: If they fail they are forced to abandon their market presence, whereas the other participants can continue no matter what happens. On the other hand the consultants are officially not even entitled to get involved in questions of coordination. Shilling's statements clarified for me what I had observed this morning in the Prime Minister's Office: Mr. K., the civil servant in charge, agreed to a deal with a Danish government advisor and a Normesian consultant. The three of them helped each other do their jobs. At the same time they made a great effort to make sure it did not look like a deal was being made. They had to disguise the fact that the initiative to coordinate the development programs could be traced back to the two foreigners.

The mediator role that is assigned to the consultant in this game compels him to intervene continually and with increasing intensity into the matters of the local administrative elite, that is, to do just what we have been doing the last two days. If, unlike this morning in the Prime Minister's Office, no intelligently concealed deals are concluded, then a specific defense mechanism will set in, like yesterday morning in the water ministry. Shilling told me how he had conducted a project years ago with the Ruritanian railway company, which offered a perfect example of this problematic. Not only were the astounding number of projects conducted within the railway company inadequately coordinated, but they actually obstructed each other. Their main problem was that the railway company lacked the steering capacity that would have been necessary for all the projects to have a positive impact. On closer inspection one could recognize, however, that the losses due to friction between the projects could hardly be blamed on mere ineptitude. Instead, the only—or at least the easiest—way for local management to maintain a minimum level of autonomy for its actions was to play the various consulting firms against each other or at least to keep them in the dark regarding connections between the projects. Shilling said that in his project he had become familiar with yet another aspect of this defense strategy, which he now saw again in the water project. He experienced time and again that Ruritanian contacts, with whom he had long been struggling—to no avail—to clarify a question, later made statements in unexpected contexts that proved that they could have contributed months earlier to efforts to resolve the problem. People preferred to appear underinformed, incompetent, or even dumb before they would act contrary to their conviction that it was generally more advantageous not to reveal everything one knows.

From a sociological perspective, Shilling had described a self-fulfilling prophecy. The defense strategies of the local executing agencies produce systematic contradictions and mishaps. These then threaten the success of the foreign consultants. Because their economic survival and careers depend on it, the consultants try to overcome these hurdles. To this end they have to intervene, using all the finesse and networking strategies they can muster. It is inevitable that their focus on their individual projects sometimes makes it more difficult for them to solve an overarching problem. It is also unavoidable that they end up reducing the sphere of action for the local participants, who in turn often react with additional, more stringent defense mechanisms and produce new contradictions in order to keep open other alternatives. Of course they often simply lose track when each project and each financier starts pulling in different directions and negotiating with different authorities in the country. As a result, the initial suspicion that the Ruritanians cannot organize their own society becomes reality. Given this contingent emergence of social patterns, the assumption that centers of calculation function as the leading agencies and are able to exert the influence attributed to them seems somewhat naive.

While taking our walk under the palm trees, we finally approached the port, where we sat on a shaded bench and could see white hovercrafts setting out for the islands not far off the coast. We relaxed with a cigarette and Shilling leafed through the study we had been allowed to take from the young American in order to copy it. He is a master of reading reports and quickly located a passage stating that in April 1996 the Ruritanian finance ministry had promised the World Bank it would present a draft amendment to the parliament by March 1999 regarding the legal, administrative general framework for running the waterworks commercially. In other words, this had occurred at precisely the same time—as Shilling stated, clearly enjoying the grotesqueness—that the three directors of the waterworks of Baridi, Mlimani, and Jamala had, in agreement with the water ministry, declared during the inception phase that amendments to the general framework had been agreed upon for July 1996. This claim was then taken up in the project contract as "milestone 1" and signed by all three parties. One could actually believe that this well-planned Ruritarian chess move had convinced the NDB to finance a project between 1996 and 1998, although it was known in Baharini that the necessary prerequisites would not be signed until 1999 at the earliest. But Shilling did not think that any one player could determine the long-term outcome of the game with such sovereignty and foresight. And on top of that he doubted that it made any sense to assume that the Ruritanian government could act in concert as "a single player." In this, as in similar cases, he was convinced that it was instead a matter of contingent entanglements resulting from what he called the general chaos in the field and what sociologists would call *distributed agency*.

This nasty discovery nevertheless prompted Shilling to take stock: One day of travel and two work days with two experts cost the company 0.2 person-months plus expenses. He said that since this effort hadn't been officially commissioned, he didn't know if he could deduct the costs. In the quarterly invoice of July 1997 he had already claimed 1.75 person-months for similar work. According to Shilling, this was typical for most projects of this kind. Although successes often depended on a considerable effort to coordinate external processes, a consultant could not officially be assigned the task of coordination as such, since this involved sovereign, local affairs.

At times it was impossible to avoid the impression that the donor organizations did not want to know all too much about this situation for yet another reason. If the coordination tasks had been realistically assessed and officially planned when the project started, this could, for diplomatic reasons, only have occurred under the rubric of the necessary contributions performed by the executing agencies. But in many cases this would put into question an optimistic prognosis for success, since even with only superficial research this indispensable contribution on the part of the executing agency could not seriously be expected. The donor would then be forced either to postpone the project or to accept substantially larger costs for project supervision, and at times even have to cancel a project completely. Once a project is approved and designed, there is very little chance it will ever be canceled. Some actors have funds to distribute; others are more interested in the fringe benefits; and yet others want to earn money. For this reason the financier prefers to turn a blind eye and calls for the autonomy and initiative of the Ruritanian partners at a time when it is actually already too late.[10] When this initiative is not forthcoming, the development bureaucracies unofficially expect the consultants to do the dirty work. That was just what we were busy doing the first two days, Shilling explained.

In the meantime we arrived at our hotel, where Martonosi was sitting on the terrace with a colleague from the University of Baharini who was looking for a job in the consulting business. They had been talking mostly about local government reforms. The University of Baharini had a significant say in such matters, but was currently being steamrollered by the speed of the events and the growing importance of commercial consulting firms. I asked Martonosi to explain more precisely what Shilling had meant this morning by the phrase "upward spiral." He replied that there were several causes for the occurrence and continuation of this mechanism that reinforce each other. First, the premodern notion of authority within the framework of modern bureaucracies shifts the responsibility to higher levels in the hierarchy. This has a dysfunctional effect in the sense that people holding a particular position are afraid they will actually have to bear the responsibility that has been formally assigned to them, since the hierarchy can stab them in the back in unpredictable ways. On top of that, responsibilities can occasionally be drawn up higher in the hierarchy in order to improve access to exploitative practices. For example, if various licenses, orders, HIV negative certificates, and the like can be obtained through bribery, it stands to reason that the authorizing signature will wander further and further up in the hierarchy.

The crucial point, however, has to do with the fact that the struggle for a balanced relationship between centralization and decentralization is part and parcel of every state administration. For instance, the decision about where a waste incineration plant will be built cannot be made solely by the local interest groups directly affected, since under those circumstances no facility would ever be built. The logic of this decision always also involves higher-level institutions. At the same time the interests of local actors must be taken into consideration at least to some degree, because otherwise they could sabotage the project. In the impoverished countries of the South this already delicate problem is connected with processes of exogenous modernization and development cooperation. This means that the higher-level, central institutions can draw more authority, power, and money to themselves simply because there is development cooperation aiming at the local level. Consequently, if central authorities manage to establish themselves as obligatory passage points for development projects, they become less dependent on the approval of peripheral people and institutions, allowing an unscrupulous political style to develop. This in turn reduces the already strained confidence in the elites of the country and in the central institutions, to the point that their actions seem fundamentally dubious and are often boycotted. As a result of this subversive reaction even more responsibility is then transferred to the central authorities, which further confirms the general mistrust toward them. Once responsibilities have been concentrated at the center, it is difficult to decentralize again. One of the primary objectives of development policies is to counteract this tendency through the principle of decentralization and deregulation. Although this approach doubtless has its raison d'être, the fact that it intensifies the main problem is easily overlooked. The administrative elite and the existing political structures are further delegitimated through externally promoted decentralization and deregulation measures.[11]

Baharini, Friday, September 5, 1997
After several hours of office work in the hotel room, we drove in the afternoon to the white sand beach in northern Baharini. While Shilling drank beer and read the *Economist* under the canopy roof of the Ocean View Hotel, I walked barefoot with Martonosi along the beach. However, after several hundred yards the uniformed guards of the hotel complex warned us about thieves who steal watches from tourists. We decided to swim back in the lukewarm ocean. After dinner on Fridays there's a discotheque at the Ocean View with a wild mix of music and a large crowd of white experts, enterprising prostitutes, student tourists, unsophisticated weekend visitors, and Baharini's *jeunesse dorée*. The flight attendants, who are always here from Friday to Tuesday waiting for a crew change, are especially exuberant dancers. I learned afterward that the prostitutes are all young and well-dressed because hotel security only admits women who fulfill these criteria—and are able to pay sufficiently large bribes. Sometime after midnight, Martonosi returned from the dance floor with a beautiful woman he apparently already knew. He introduced her to me as Eva, Miss Ruritania 1995. After drinking a Baileys, the young woman returned to her European

companion, who was waiting impatiently at their table. On the drive back, Martonosi took the wheel. He put a Tracy Chapman cassette in the tape deck, turned up the volume, and drove with obvious pleasure and great familiarity through Baharini by night. When a couple of sinister figures surrounded the car at a red light, I heard the short mechanical click of the front and back doors locking, before Martonosi floored the car through the intersection.

5   Trading Zones

Preliminary Remarks : Edward B. Drotlevski

Jamala, Saturday and Sunday, September 6–7, 1997

Beyond the slow-moving traffic of Baharini, the asphalt street on the other side of the river improves and everyone steps on the gas to make up for lost time. Policewomen in pristine white uniforms lurk behind bushes with state-of-the-art laser technology, on the lookout for traffic offenders. But we were lucky and made it through. At a major crossroads we picked up some skewered grilled meat and observed several sweat-soaked people getting out of overloaded minibuses. Most of the women were carrying bulky bundles on their heads, with an infant on their backs and a toddler in tow. Armed with hiking sticks, their husbands set out in a stately manner ahead of them, their sights set on the mountains along the horizon. After the crossroads the street runs along a river back down to the ocean. Sisal fields, most of which are no longer tilled, extend for miles, as far as the eye can see.

Most of the industrial facilities along the access road to Jamala are out of operation, with bushes growing out of the windows and rain gutters. The abandoned marshaling yard at the outskirts of the city is evidence of an era long past, with its trackage, steam locomotives, and maintenance buildings. Departure and arrival times of trains no longer running are posted on the frontage of the passenger station. Our first stop was at an obelisk commemorating the heroic death of colonial soldiers. Sitting on the steps of the memorial in the evening sun, we had a cigarette and enjoyed the view out to the ocean. Directly in front of us was the nonfunctioning port encircled by an entire fleet of rusting ships that mark the end of modern seafaring for Jamala. Right in the middle a dhow from Yemen was being unloaded manually; it was the kind of sailboat that Arab traders used 1,500 years ago along the coast of East Africa.

We picked up the key to our lodgings from the project office at the waterworks. The newly renovated administration buildings are located on the grounds where the colonists set up the first groundwater pump in 1908. At the time the grounds were at the edge of town, but now they are part of the city center, where the groundwater is contaminated. This was one of the reasons why a new water extraction and conveyance facility was built with NEB support at the river far outside the city. The S&P offices are in a formerly affluent residential area on a hill at the outskirts of town. The high palm trees in the well-tended garden, the double-gated entrance, the spacious house, and the exotic wood furnishings are reminiscent of a different age. Today the water boiler on the roof was empty and the refrigerator was warm, since there had been hardly any electricity in the past twenty-four hours. For dinner we drove to the only good restaurant in town.

The guarded parking lot there was filled with typical project cars: new Japanese four-wheel-drive vehicles with the names of the projects on the front doors. Our silver Toyota was labeled "Baridi Water Supply Project." The inscriptions on the other cars documented the wide range of organizations and countries active here, as well as their chosen problem areas: fighting AIDS, combating coastal erosion, water resources management, railroad rehabilitation, port rehabilitation, promotion of village development, street maintenance, forest cultivation, the coconut industry, and so on. Only one table in the restaurant was occupied by a Ruritanian family. Project people—that is, nonnatives who are not tourists—were at all the other tables. The longest table was reserved for a Christian community that had invited an Evangelist from Nigeria to be its guest; they had held a conference in the afternoon on "Economy and Faith."

The waiters welcomed my two companions and offered them their favorite dishes. Thankfully I learned that the small imported bottles of beer are preferable to the large Ruritanian ones, since one can polish them off before they get too warm from the hot air. We were dripping with sweat as we drove to the hotel at the port after dinner. A cool evening breeze turned the hotel garden into the most pleasant, comfortable place in the city, where you could sit in moldering wooden armchairs around wobbly wooden tables sipping whiskey on ice and gazing at the starry tropical sky. The lawn, which used to be luscious green and immaculately groomed in English style—as Martonosi remarked—has unfortunately been burned by the sun. This was one of the visible consequences of the project in the city. The hotel has been made to pay its bills since 1996, as is right and proper, so it stopped watering the lawn. Because today was Saturday, the sounds of disco music wafted over in the wind and a couple women and men were still looking for escorts. The routine program ended with a second whiskey and the drive back to the residence, where the night watchman had already fallen sound asleep.

On Sunday afternoon, September 7, the team leader T., the project engineer I., and the IT expert S. arrived from Mlimani and Baridi in two additional Toyotas to attend a project team meeting. Of the six experts, the only one missing was B., who is responsible for the waterworks' financial accounting.

At the Project

Jamala, Monday, September 8, 1997
In the team talks, which extended over the entire day, references were continually made to things I had never heard of. And there were a lot of allusions to the background of the project. Only gradually did I start to comprehend what it was all about. The first team leader had been project engineer I. He was succeeded at the end of 1996, after only a few months, by T. The first team leader had not been able to coordinate the activities of the six experts in different fields that were located at three separate sites and interrelated in terms of both substance and timing. There had been a falling out and they all started passing the buck back and forth. In the end, the executing agents—that is, the three UWEs (urban water engineers) of Baridi, Mlimani, and Jamala—had the impression that the project had too many cooks and no unified plan.

The conflicts seemed to have focused on two main issues in the course of 1997. First, the engineer had been put in the difficult position of having to solve technical tasks for an Organizational Improvement Program (OIP) without the necessary means at his disposal. This required the much-invoked improvisational artistry of finding localized technical solutions with minimal means under adverse conditions, which at the same time were supposed to strengthen the organizational competence of the waterworks. However, the three UWEs felt they had mastered the necessary survival skills better than anyone else—and they certainly didn't need the help of any highly paid Normesian engineer for this. Consequently they did not include him in their everyday fiddling to keep the technical operations going. To prove his importance the engineer saw no alternative but to constantly discover new weak points and then to call—in good Normesian engineering style—for comprehensive technical solutions. He put the case for optimum high-tech solutions to the UWEs, who wanted more than anything to have the entire technical facility renovated through the Normesian development project. At the same time, this demonstrated his own superfluousness in the OIP, which is not responsible for this kind of solution.

The second point was to reform the system of process control. This aimed for increased reliability of processes by transferring human competence to computers. The software expert S. was the only one who really understood this key aspect of the project. To all the other team members, turning the entire measure into a computer program meant working toward something that they ultimately could not grasp. Obligatory passage points always appear threatening and this was exacerbated in our case, because a customized program had been selected in order to take local particularities into account. If S. were to throw in the towel it would be difficult for S&P to provide the system, which was not yet fully installed and did not yet run properly, for the same cost and by the agreed deadline. It is common when using customized software that problems arise here and there in the first trial runs with real data. In our case these small, normal errors added fuel to rumors that S. and his water management system (WMS) were plagued by larger problems. Shilling was aware of the undermining effects that such talk could have, so it was important to him to set things straight whenever he could.

Although the data provided by the UWEs was incomplete and inconsistent, the project could not wait any longer for more reliable data. Consequently, S. had to keep adapting the WMS. At first the program was so conceived that on a designated day the complete data record would be imported into the system through the only "door" provided for that purpose, which would be closed off afterward. This would ensure that there would be a single master database that could never be altered. That in turn would mean that all subsequent changes could be saved only in certain secondary files, which could likewise not be overwritten. Every new modification of a record would register the date and user's code number. Thus it would be possible to monitor the differences between the original entry, the first new entry, and all subsequent new entries, thereby preventing any user of the program from manipulating records surreptitiously. Because the data quality necessary for a major, one-time import was not available, S. had to deactivate the program's complex protective procedures and allow individual records to be imported at any time. This required him also to refrain from installing the security measures that would prevent data from being overwritten. According to the original program design, continual importation would have led to an inordinate number of secondary files that registered every correction to a correction to a correction ad infinitum, which would have caused the program ultimately to shut itself down. Thus one of the major aims of the project—to have the computer prevent data fraud and confusion—became nothing more than a farce, hampering the ambitions of IT specialist S.

From what the project people are saying, there is no doubt that the WMS is a main protagonist in the project and a full member in an actor network, equipped with human qualities. It has its good days and bad days, it has its teething problems, it is S.'s baby, and ultimately it is supposed to accomplish everything. This also means, however, that if the project goes awry it is because of the WMS. The WMS is the manifestation of the entire project idea. Since S. is the only one who has any control over the main protagonist in this drama, it is not surprising that he is quickly made into either the savior or the scapegoat of the whole thing. The scapegoat function becomes particularly precarious in combination with the project's main problem: the interface between data responsibility and software responsibility. As of June 1997, most of the WMS's teething problems could be traced to the fact that the expert was required to perform unplanned modifications in order to make the system compatible with the data. The UWEs, on the other hand, claimed that their data were not faulty, but rather that the WMS was incompatible with the data. This line of argumentation made S. furious; he insisted: "My Water Management System imports any random data format as long as the data records are consistent." Martonosi said that some fundamental action had to be taken at this point or else everything would go wrong in the end. He wanted to try to solve a communication problem, not a problem of factual issues. His first question was: "How can we manage even to agree on what the factual issue is?"

After the team went through the engineering and IT areas of the project, it finally moved on to financial accounting. B. is in charge of introducing a business accounting system. Compared with the other project areas, this appeared to be running smoothly. As everywhere, however, there were delays due to the fact that the computer networks went into operation about six months later than scheduled, the training program was taking longer than expected, and the generation of lists with the fixed and current assets for the opening balance was dragging. The crux of the matter here, similar to the situation with the WMS, was the outdated, incomplete, and inconsistent data. After having to spend considerable time gathering the data, the team will have to import them with all their gaps and errors, and bank on the data being corrected over the coming years.

A crucial point regarding the data lists is more obvious in the case of accounting than in that of customer localization: If the existing data dilemma cannot be resolved, it is not possible for the accountants simply to start from scratch, because there are contractually stipulated claims and debts. Ultimately, this would lead to publicly admitting total failure, something which no organization striving for effectiveness in the future can afford to do. S. became so desperate that at the end of the ten-hour discussion he suggested arson as the way out. That would be the only way to escape the problems of the past, save face, and start again with a clean slate. Thanks to the disciplining effect of his software, the staff would only be able to enter consistent data records in the future—regardless of how they felt about it.

The Triviality of Lists

Jamala, Tuesday, September 9, 1997
Yesterday's project discussion focused mainly on the question of customer data. Basically, as S. explained to me, everything is very simple, and you have to constantly remind yourself of that to avoid going crazy. Administering and operating a water tap—in economic terms, a "point of sale" (POS)—requires six data records and reliable cross-referencing of those six records.

(1) Point-of-sale account number
(2) Localization of the water tap
(3) Name of the customer assigned to this account number and who pays the bill
(4) Customer's mailing address
(5) Number on the water meter and, if applicable, average calculation rate
(6) Customer's account balance on the day it is entered into the new system
(7) Reliable correlation of (2)–(6) with (1)

This list is set up logically with the numbers of the points of sale at the center (1). Of course, a point of sale only remains a point of sale as long as it can be reliably located in the supply area (2). It is otherwise an illegal tap through which an inestimable amount of water is lost without any possible recourse. The connection between (1) and (2) is manifested through a reliable correlation between a simple list of customer account numbers and dots on a map. The remaining four points of information, (3)–(6), and their reliable correlation to (1) are absolutely essential in administering an existing point of sale. If one of these is missing or incorrect, the procedure will not work.

S. was convinced that the trivial logic of this system was clear from the outset to everyone involved. The UWEs could not delegate the correction of customer data to a highly paid Western consultant precisely because the logic is so compelling. To the extent that S. held to this interpretation, the question of why the customer data had not yet been corrected must either remain incomprehensible to him or he had to look for other explanations. He actually had the nagging suspicion that he was being given the runaround and that in reality something totally different was at stake. S.'s further comments on the data dilemma, however, made me think that his suspicion that the waterworks staff was conspiring against him was misguided in a revealing way.

S. regularly discovered all kinds of errors. Random sampling has shown that an astounding number of even elementary lists are *invalid*. S. told me that elementary lists contain data (mostly numerical) on a first-order representational level. These are figures that do not refer to other, preceding numerical representations, but are instead related directly to a countable reality, such as a concrete number of water meters. Most figures used by people in development organizations are located at higher representational levels and are consequently aggregates of figures from the lower representational levels—such as the notorious collection efficiency. But the lowermost end is always composed of first-order figures, or elementary data.

First-order figures are invalid if they do not coincide with reality. It might say on paper that 750 of 1000 existing water meters function properly, but S. cannot count on there being 750 that really function. The question "Which meters function properly?" moves up to the next level of aggregation, which involves a second list. In order to maintain its identity as a water meter, each meter must have a number engraved on it and that number must be entered onto a list. If someone wants to know where the broken water meters are, a third list is then used, which itself is an aggregate. The overall validity is further reduced when these three lists, each containing not entirely valid data, are combined. The end result is that in Jamala it is virtually impossible to proceed from the list of water meters to the actual site where a water meter with a particular number is located. According to S. this leads to the next systematic weak point.

It has turned out that most of the lists located above the first-order representational level are not sufficiently *reliable*. The existing procedures for aggregating series of figures and for linking the lists are not followed on a regular basis. Within one and the same list, therefore, individual figures have often been generated according to different formulas, which are impossible to reconstruct after the fact. Informal corrections are frequently made, which might be based on what a staff member happens to remember, and the number itself can no longer be verified by sight. Employees tend to assign greater significance to individual cases and their own memory than to a system or the organization's memory. For these reasons, the unreliable list can no longer be reviewed and corrected through comparison with other lists.

Random sampling has shown that lists made up of data consolidated from other lists have individual data records at varying states of completeness and in various formats. The lists are therefore *inconsistent*. The causes for this are again insufficient commitment to the system and a general lack of system standardization.

The waterworks of Baridi and Jamala started correcting and completing their customer data in 1994. This major action to gather primary data was called a customer survey. Here in Jamala, S. had a closer look at the results of the survey and how it was conducted, because he had been drawn into completing this project although contractually it was not within his area of responsibility. To illustrate this problem to me, he brought me to a room in the Jamala waterworks specifically designated for the work of the customer survey team.

Using files and maps, S. explained to me that despite claims by the management to the contrary, data collection was in fact interrupted before the entire supply area had been reviewed and before the numerous gaps in the previously reviewed neighborhoods could be filled. As he was presenting the procedure to me, S. gradually worked himself into a state of perplexed amazement, as if he were getting an idea of the magnitude of the problem for the first time. The supply area had been divided into twenty-three maps sometime in 1994, before the operation began. The boundaries of the individual maps sensibly coincided with the borders of the nine technical zones of the network and the maps were named according to the city areas they represented. On the customer forms that the data collection team had to fill out, there was one category called "area." Everything seemed well prepared up to that point. During the data collection, however, team members entered onto the preprinted forms the commonly known neighborhood name that customers mentioned, instead of the map name. These neighborhoods do not have distinct names—each one can be named and measured differently depending on context and speaker—nor do they coincide entirely with the nine technical zones or the twenty-three maps. Consequently, they ended up with 110 different names for "areas."[1]

The reason for collecting the data was to make customers relocatable. Physical points on the ground—the individual taps—were supposed to be correlated with representations of these points on a map and on a list so that it would be possible at any time to find a point on the ground using its representation on the map. This would facilitate tap maintenance, for instance, as well as bill collection. At the end of this operation, however, the respective map name was not listed on the individual customer forms, and the customer numbers were not recorded on the individual maps. To top off S.'s amazement, these errors were not even noticed, but virtually all customer forms along with the nonsensical entry under "area" had all been entered into a computer file.

There was yet another procedural error. In Jamala, house numbers do not exist as a basic principle to aid mail delivery. The National Housing Corporation (NHC), as the former (socialist) owner of most of the properties, simply painted its own numbering system onto the houses. This system was also used to collect customer data for the waterworks. But the homes were gradually returned to their former owners, who painted over the numbers. As if that were not enough, during the data collection, the team often divided into two groups, each one setting off from opposite ends of a street for their research. When they came to houses without any NHC numbering the two groups frequently assigned their own numbers, starting in each case with 1. This resulted in the lists for many streets having two houses with the same number. To top off the chaos, the completed customer forms were all mixed together in the office, so that even the route taken by the team could not be used to help clear up any confusion.

Finally, the data that were collected in 1994 were not entered into the master database used for billing until October 1996 and later. After the data collection of 1994, new information was subsequently entered only into the master database that was current at the time, but not into the new database. As a result, when the master database was "corrected" in 1996 with the data collection of 1994, in countless cases correct data were substituted with incorrect data whenever, for example, the water meters had been replaced or the customer had changed. The IT expert is certain that such a fiasco cannot be blamed on a lack of knowledge or on cultural differences. Instead he presumes the causes are perhaps disinterest and carelessness, yet he suspects first and foremost a conspiracy against him.

In contrast with this interpretation, I am astounded by the suggestion that the IT expert or anyone else on the project could even imagine that merely re-collecting the six primary data needed to identify a customer would solve the problem of notoriously unreliable data. The recent attempt to collect the data anew under the name "customer survey" seemed to prove to me that the problem is a far-reaching, culture-specific failure to recognize the *priority of the process* over the elementary data and over the attempt to be accurate in individual cases within the database. Put simply, data collection was a fiasco because it was not based on any overarching, well-thought-out process. Because this was precisely the reason why the new collection of data had been considered in the first place, it should have been obvious where the priority needed to have been placed—on improving the process and not on correcting the data.

Another observation also puzzled me. Our conversation took place in the customer survey team's room. It was here that S. worked with the lowest levels in the hierarchy, which had been assigned this supposedly trivial task, but obviously did not understand it. Afterward we went together to Joseph Mutahaba, the UWE of Jamala, whose air-conditioned room was only about twenty yards away. Mutahaba's critical questions kept putting S. into a situation of having to verify or explain something that was going on in the customer survey team's room just opposite. Mutahaba viewed the circumstances presented by S. to be so trivial that he could hardly believe they were causing problems. He presumed instead that S. did not properly understand the members of the customer survey team, since they spoke only rudimentary English. In the end the young engineer C. was called in—he was in the second level of the hierarchy and responsible for data collection. His report on the latest developments in the process of data correction did not contain any of the problems that S. had described a short time earlier.

In other words, the IT expert of the project and the head of the waterworks, who was the executing agent, had not been able to communicate—for a year already—on what was going on in the next room. The option of simply walking together over to the room in question apparently did not occur to anyone. Nevertheless, I believe that my reflections here have finally brought me to the empirical level necessary for dealing successfully with the somewhat mysterious phenomenon of list autophagy.

Mlimani, Thursday, September 11, 1997

Yesterday Shilling drove to Baharini in order to fly to another project. After the final meetings in the waterworks of Jamala, the other project workers all set off for Mlimani. For those who remained, our visit must have seemed like a hurricane, after which calm finally sets in again.

We moved along quickly on the new, rarely used street. At the roadside we again saw broken-down buses and trucks. Martonosi explained how this street had turned into a downright death strip after it was repaired a few years ago. The police had since imposed a ban on night driving and they also enforced the speed limit more effectively than they had before. In the villages the road ran through, improvised speed bumps had been constructed. On a longer downhill segment of the road we were passed by one of those notorious buses, although we were driving 75 mph. On the side of the bus we could read its purpose and destination: "Video Coach, Baharini–Baridi"; and the driver's motto, to be as fast as a fax, was proudly printed on the back of his bus: "Fax from USA." Ruritanian buses look exactly like the long-distance buses of many poorer countries of the southern hemisphere: The windows are darkly tinted with curtains drawn behind them. The passengers either sleep in their seats or try to follow a Kung Fu film on the small screen at the front. The deafening roar of the Japanese diesel motor is drowned out by a stereo. European passengers, usually young students, border on panic from the noise, the claustrophobic crowdedness, and the insane speed. This is exacerbated by the fact that the Ruritanian passengers are apparently perfectly comfortable with the situation. Ironically, this gives rise to the very sense of unbridgeable foreignness that traveling with the cheapest means of transportation is supposed to overcome. We had a better chance of avoiding this unpleasant experience from the distance of our Toyota, in which we were listening to a Schubert tape.

We arrived in Mlimani before sundown. The project headquarters are located here: S&P rented three residences for the long-term experts and their families. Since nothing remained of the aura of the Grand Hotel of Mlimani, we preferred accommodations in the church-run hotel management school for orphaned girls. That was practical insofar as the UWEs used the school's assembly rooms for their meetings. The school director is also head of the board of the Mlimani waterworks, and, as a member of parliament, she is influential everywhere. Martonosi informed me that this is where the strategic planning workshop that we had talked about had taken place in October 1996. In the meantime I had developed an eye for water issues, so I immediately noticed the suspiciously green hotel grounds. Later

I learned that that was in fact a long story that had never been completely resolved. But now there was a water meter and they did pay their bills; only the old debts were still disputed.

The Cunning of Lists

Mlimani, Sunday, September 14, 1997

There is an out-of-use kitchenette behind an office building on the grounds of the waterworks of Mlimani. After wiping away the cobwebs on the wooden shelves it was possible to identify cartons that had been filled with papers a long time ago. Martonosi pulled out a sheet to have a closer look at it in the office (see figures 5.1 and 5.2). It was an 11x17-inch file card intended for use as a house payment record for a water customer. A few features immediately caught my eye.

In the upper left, three of the six spaces for elementary data relating to the customer had been left blank. The street name, street number, and zone of the pipe system had been filled in, but not the name of the customer,

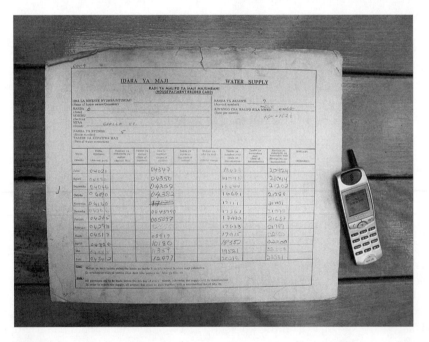

Figure 5.1
Customer record card (front).

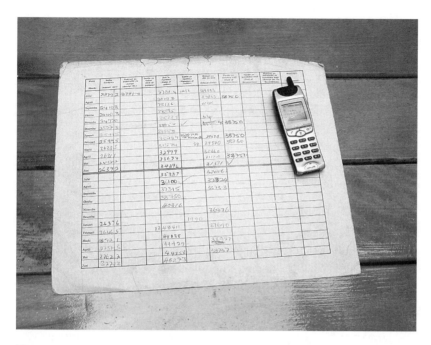

Figure 5.2
Customer record card (back).

the section within the zone, or the date the tap had originally been connected. In the upper right, the account number had been recorded as a "9," which does not jibe with the customer's assignment to zone 0. The correct account number, 0009, appears in the upper left, outside of the printed border. Below the space in the upper right where the monthly rate should have been recorded, there is an "N/M–1026" entered below a crossed-out number. This denotes that a new meter (N/M) with the recorded number had been installed. The lower half of the card indicates that it was designed to keep record of item number 6 of the six elementary data sets necessary to manage a point of sale, as IT expert S. had explained to me in Jamala. This part of the card shows twelve rows for the twelve months of the year starting with July and ending with June (Ruritania's fiscal year begins in July). For each month there are eleven columns: the first lists the name of the month, followed by columns for "amount paid," "receipt number," "date of payment," "name of cashier," and so on. However, the card was not filled in according to the predefined headings; it had been used as a chart with columns to record the meter readings following the months listed in the first column. In order to save paper, the fifth, eighth, and tenth columns were used to record the meter readings for the subsequent three years. From what one can see on the front of the card it is impossible to know which four calendar years were recorded here.

On the back of the card (figure 5.2), the meter readings for another seven years were recorded following the same pattern, whereby twenty-five of the eighty-four months have no readings at all. Three calendar years ('88, then again 1988, '89, and 1990) were entered in between the described columns, and after some guesswork one can figure out that they refer to the columns to their right. It thus becomes possible to count backward to attribute a calendar year to each column so that the card seems to cover the period from July 1980 to June 1991. The date the new meter (number 1026) was installed (March 7, 1989) was recorded in the space for February 1989. According to the figures entered, the customer's annual water consumption varied enormously, from 321 units (July 1980 to June 1981) to 8,130 (July 1981 to June 1982) to 9,936 (July 1989 to June 1990), with three years around 6,000, interrupted by another three years around 2,000 units. Between June 1985 and January 1986 apparently only 2 units were consumed. According to the last column, from August 1990 onward the meter presumably stopped registering any water consumption at all, although readings were still sporadically taken until May 1991, when the procedure was discontinued entirely.

The main problem with this kind of record keeping is not that some of the information is lacking, some is incorrect, and none of it was entered onto the correct form. Most of this could be reconstructed and corrected with considerable time investments, just as one can figure out that the first column must refer to July 1980 to June 1981. The main problem is instead that all the elementary data of the waterworks are available only in this form; aggregating this kind of poor data inevitably leads to the systematic production of new errors, which escalate from one step to the next. The card evidently came from one of the many previous attempts to establish ever new bookkeeping systems in the Mlimani waterworks in order to avoid precisely this spiral of escalating error. These systems might have been designed poorly in one respect or another, but that cannot be the reason why they all fizzled out after a few years in use. In other organizations, design errors are gradually corrected over the course of time.

Getting to the bottom of this problematic development requires a closer look at the connection between various printed forms and lists. For example, the numbers on the form shown obviously came from the route list of a meter reader who went from customer to customer to read their meters. After the numbers of the route list were entered on the customer form discussed here, the difference between the old and new meter readings was calculated and entered onto a separate list. Following this bureaucratic process further, a complex system transpires in which figures from one form were transferred to another, frequently changing its aggregate structure. In many instances, they were combined with other figures and then had to be incorporated into two separate cycles to allow the billing and accounting departments to monitor each other. This is all part of what necessarily has to happen in any rational bureaucratic organization that requires accountability. However, during the inception phase of the project, Shilling and Martonosi found out in Mlimani that any connection between billing and accounting had virtually petered out. When preparing the invoices, the billing section stopped making carbon copies that normally went to the accounting section to make it possible to check how much of the billed amount had actually been paid.

In the process of solving this problem, it also became apparent that the complete customer list of the billing section was based on the route lists of the meter readers. This improvisation became necessary because at some point the old customer lists were largely obsolete. In this case the degeneration process had become so exacerbated because the lists were recorded by the IT department of the ministry in Baharini, which ignored corrections sent from Mlimani. Consequently, the locally and centrally recorded lists continued to diverge until the latter were finally declared to be worthless. The solution that was chosen is very interesting. Since the meter readers are the only ones outside of the office who have regular contact with the real situation of the customers, they are the ones who best know how many customers there are in their streets, what their names are, where exactly they live, whether or not they have a water meter and whether it functions properly, and finally, what the reading on the meter is. Therefore, the route lists of all meter readers were compiled and the results were declared to be the new customer list. This, however, opened up the possibility of meter readers being bribed to exclude someone from the list.

There was no way to refute the impression that a certain procedure had favorably merged here with a certain instrumental logic of maximizing benefits. I believe, however, that this supposition is based on a false premise. It is assumed that the people involved understand the system of documentation, but I think the examples mentioned tend to prove the opposite. I also do not find it very convincing to attribute such events to the staff's education level. The degeneration of a contrived system of organization must be the responsibility of the management, which certainly does not suffer from a lack of education.

The entire mechanism suffers from what Martonosi and Shilling call *list autophagy*. Specifically, they mean that when a new system of documentation is installed and put into operation in the waterworks, it consumes itself little by little. With each new generation of figures that are circulated, the lists gradually lose their validity, until they finally become worthless. The cycle ends with the staff proving in a kind of self-fulfilling prophecy that the procedure does not make sense. The individual memories of the workers and the collective memory of informal networks offer far more in the three waterworks than the institutionalized memory of the documentation system. Someone who needs information and looks through the existing records for the answer demonstrates nothing but his or her own naïveté. Rather than being so credulous this person should find out which staff members have the information in their heads.

Nevertheless, the project is based on the assumption that participants can be expected to deal appropriately with written documentation and figures, files and procedures. For this reason I think the following question has to be posed: Could it be that these problems are caused by a rejection of—or perhaps a culturally motivated nonadherence to—the fundamental way that bureaucratic systems function? Martonosi hates this question. He always shifts it back to the level of power-saturated hegemonic negotiation processes that prevent local knowledge from having any chance of becoming the foundation for the new administrative systems. This, he insists, is why the forms are not adapted for the local context.

Within the framework of the technical game, it appears legitimate to attribute the unreliability of the lists to the lack of material stimulus for better performance, to deficient training, or even to paternalism, but it is by no means legitimate to take into account a cultural difference. On the quiet, however, everyone seems to agree that this difference does in fact exist. The team leader says, "It will take generations." The engineer sees no prospects at all; Shilling is accustomed to sarcastic, calculated optimism; the IT expert seeks refuge in his conspiracy theories; the UWEs put their hope in the new computers; and Martonosi elegantly disregards the problem insofar as he believes the basic hegemonic structure of the negotiation process is responsible for the fact that local forms of knowledge are never even given a chance. In doing so he is observing one of the strictest taboos of contemporary anthropology, which absolutely prohibits using people's cultural patterns to interpret their actions.[2]

The Secret of Lists

Mlimani, Monday, September 15, 1997

Today I'm having a desk day, so I can think through my notes on the subject of "lists." First I'll review the comments I wrote down while I was still in Urbania:

A *list* is a record of things or abstract statements that have been removed from their context and written down one after another as *facts*. The classification system and selection principle according to which the facts in a list are chosen is not included in the list itself. Usually, the principle used in ordering the facts in the list is linked neither to the original context of the facts nor to their selection principle, but instead to the logic of the list itself, such as using the alphabet, a series of figures, or the date as the criterion of findability. Several lists linked together yield a *table*. A table is made up of intersections of lists in horizontal rows with lists in vertical columns, which allow statements to be made that go beyond the scope of the individual constituent lists. If the points in a table refer not to objects but to a sequence of systematically interrelated actions, the result is often formulas or *recipes*. In the case of the project, for example, the action plan is one of the recipes referred to here. Another type of link that merges two lists creates a causal relationship between them. One list contains a series of symptoms—with respect to this project, for instance, the list of defects determined by the feasibility study; the other list contains a series of causes for each phenomenon—here, the list of solutions from the feasibility study that are linked to the causes. The shorter the list of causes in comparison with the list of symptoms, the stronger the explanation.[3]

These observations seem trivial at first glance. This is because the classification and selection work that precedes every list has successfully been rendered invisible. Using the language of Mary Douglas, the classification system that the list is based on has become so well institutionalized that it is erroneously viewed as being a characteristic of the things themselves. Not until a list gets into the wrong hands, so to speak, does it become clear how many presuppositions are necessary in order to understand it. This fact is illustrated most clearly by the taxonomy attributed "to a certain Chinese encyclopedia," (which is why Borges took it up and Foucault cited it) according to which animals can be divided into the following categories:

"(a) those that belong to the Emperor, (b) embalmed ones, (c) those that are trained, (d) suckling pigs, (e) mermaids, (f) fabulous ones, (g) stray dogs, (h) those that are included in this classification, (i) those that tremble as if they were mad, (j) innumerable ones, (k) those drawn with a very fine camel's hair brush, (l) others, (m) those that have just broken a flower vase, (n) those that resemble flies from a distance." Many lists that seem completely natural and logical to their familiar users would arouse similar amazement among distant observers if they were far enough away (hence the "Chinese"), because "obviously there is no classification of the universe that is not arbitrary and conjectural. The reason is very simple: we do not know what the universe is."[4]

Thinking in lists or in the facts and figures contained therein is also based on a second form of invisibilization. Facts that have been removed from their context and appear unrelated in a list have no meaningful connection to a story or image. They are therefore difficult to remember and need to be put in a written form as a substitute for the narrative context. List knowledge is consequently always written knowledge. As far as our present knowledge of cultural history goes, writing first emerged in the area of bookkeeping as a means of storing data. The first specimens of writing were lists of tax revenues, of claims and debts, of information on products' places of origin. Writing did not develop and spread as a means of communicating narratives, but as a means of data storage, as the exteriorization of memory. This distinction gives the impression that narrative knowledge belongs to everyday life and to the domain of interpretive, speculative knowledge. In contrast, science and rational organization have struggled to keep their distance from all narrative knowledge, from myths and legends. Officially, they have to remain untouched by it and be grounded solely in fact; otherwise they risk losing their legitimacy. This operates according to the following dichotomy:

Objective/written/list form ↔ fictional/oral/narrative

Recent research, however, has shown that organizations do not in fact devalue all narrative knowledge, nor do they even attempt to do so. Put succinctly: Shelving things, putting them *ad acta*, is a form of systematic forgetting of written memories that provides narrative knowledge or oral memories with the necessary space to unfold. Of course, in a society thoroughly regulated by putting things into writing or even into law, written records assume enormous significance in the case of a conflict, as opposed to smoothly running, routine operations. Written agreements, contracts, and all kinds of documents are drawn up, at the very latest, in court. But here too the inevitable gaps are once again bridged through interpretations, narrative knowledge, and people putting their foot down. Knowing about the existence and potential significance of extensive written records and the exteriorized memory they contain thus plays an important role in the construction of oral narratives and their relationship to power. This means that within the context of formal, rational organizations, which in case of doubt or emergency refer to written records, the prevailing oral tradition is of necessity completely different than it is in societies lacking a system of writing.[5]

That is everything from my notes on the secondary literature, but what does this mean for our actual case? First it is important to keep the practical goal in sight. The question for the waterworks is how to get a translation chain to work between two very remote translations, A and Z. A is the small step from the water flow in the pipe to the turning of the cogwheels of the water meter and the subsequent meter reading. Z is the last translation in the chain: the step of the customer in the payment department of the waterworks, where he or she deposits payment for services and the cashier correctly books the transaction, the ringing of the tills announcing the successful transubstantiation of water into money. Half the job is getting the path from A to Z to function properly. After payment at point Z, it has to go through practically all departments and steps that keep the process going to then return to point A, the pipe end as the point of sale (POS) where water should continue to flow, be measured by a meter, and, again, transformed into money. For this cycle to function, people, technical equipment, and documentation must be intelligently interwoven. In management terms, the process must be controlled. Of the diverse problems posed in this area in the waterworks of Baridi, Mlimani, and Jamala, the trickiest is maintaining the translation chain through bureaucratic representation.

According to observations up to now and preliminary theoretical considerations, the issue here is the self-evidence of bureaucratic representations or the relationship between counting numbers (written knowledge in list form) and recounting stories (oral knowledge in a narrative form). It is possible to distinguish three levels:

(1) Elementary facts and figures, for which a simple criterion of objectivity taken from correspondence theory is adequate. There are, say, 1,000 or 3,000 water meters.

(2) Elementary procedures to aggregate facts and figures. An elementary procedure is either logically true or false. Someone who assumes that daily

water consumption can be calculated from the daily amounts of water produced is simply mistaken. Someone who bases his or her calculations on the amount of water billed is on the right track, if methods are found to register and record the amount of water that is lost and that which is consumed but not paid for.

(3) Strategic procedures related to how a specific reality is defined, what goal is sought, and what consensus on these questions exists among experts. Here it is not possible to distinguish between true and false, but only between effective and ineffective, or between just and unjust. Anyone who conceives and evaluates a project about increasing collection efficiency has introduced a strategic procedure that is oriented around an external goal or purpose, and not around the matter itself.

These distinctions, however, are part of a heuristic process that makes sense and is helpful only as long as it is used as such and not absolutized. In reality, namely, there is also a causal logic between the three different levels that proceeds in the reverse direction, that is, from (3) to (1). According to this logic, there is always initially a paradigm (3); then the related elementary procedures are determined (2); and finally, the elementary facts and figures are produced via these procedures (1). If something changes at the paradigm level it is passed all the way down so that different elementary facts and figures become significant or the old ones are given new meaning. This does not mean that no new facts can come to light within the framework of existing paradigms. However, we should bear in mind that the one-sided logic proceeding from (1) to (3) presumed above systematically suppresses the retroaction from (3) to (1).[6]

This suppression determines the specifics of bureaucratic systems. Everyone knows these specifics from their own experience. For example, suppose someone raises a claim against a bureaucracy; even if the objectivity of that claim can be easily confirmed, it is still possible for the bureaucracy to rightfully reject the claim if no category exists for this particular case. In the eyes of the bureaucracy the said case does not even exist. This might drive the people involved to desperation, but the rationality, effectiveness, and legitimacy of the bureaucracy is based on stubborn adherence to procedures. What counts here is not objectivity that corresponds to an external reality, but objectivity that corresponds to a procedure.

Within the scope of the Organizational Improvement Program, which is the reason I am sitting here in Mlimani, they are basically concerned with introducing the *principle of procedural objectivity*. In material terms this means that the project concentrates on the water management system (WMS), which is supposed to determine what data are relevant, how they are to be formatted in order to be recognized as data, and how these data are to be linked together to constitute information. Within the framework of the technical game, ostensibly neutral, elementary procedures will be introduced, which are supposed to correspond to an unproblematic reality of facts and data. The question of procedural objectivity is not even raised in the official presentations, nor is it mentioned in the ongoing negotiations. This serves to mask what are evidently the most delicate points, on which the success of the project largely depends. Martonosi, the project anthropologist, actively contributes to this avoidance for reasons of his own.

In the first place, this avoidance strategy ignores the fact that the principle of procedural objectivity invariably makes unreasonable demands on common sense. In Ruritania, these demands simply exceed the tolerable limits set by the prevailing commonsense understanding of objectivity as fidelity to the individual case at stake. People are not willing to adjust their sense of reality to a mysterious procedure that subordinates the complex and always specific reality of individual cases to the "reality" defined by the predetermined categories of a printed form. This resilience comes as no surprise if one bears in mind that the institutional order of Ruritania and its colonial and postcolonial legacy never really demonstrated their superiority in providing justice and prosperity.[7]

The avoidance strategy also ignores the fact that the principle of procedural objectivity is based on a relationship between written and oral traditions that evolved over time and is socioculturally embedded. In Ruritania there is evidently neither a great need nor sufficient opportunities to give priority to written documents over narrative knowledge. In neither of these two aspects—that is, the priority of procedure over substance and of counting numbers over recounting stories—has the point been reached at which a procedural order becomes legitimate. Bureaucracy remains unable to establish any desirable configuration powerful and successful enough to supersede individual injustices and inconsistencies inevitably produced by its standardized procedures.

Up to this point my interpretation agreed with that of Martonosi. But our opinions diverged over the causes of and the options for overcoming this dilemma. For Martonosi, it is the hegemonic constellation that causes the technical game to mask the main problem. He believes that the models to be transferred have been given a nimbus of neutrality and objectivity in order to conceal their colonizing power. This gives rise to the following equation: In the same way that "2 + 2 = 4" is true everywhere and always, the sentence: "Whoever wants to operate a waterworks has to do the same things no matter where, when, or by whom it is done" is also supposed to be universally valid. In reality, however, the technical game functions like a Trojan horse in the sense that it smuggles in a new social order and a new web of belief. Talk of elementary data and procedures continues until everyone forgets that this involves a fundamental intervention in the production of an urban order, for which there are many alternatives. These possible alternatives, which would introduce prospects for a successful adaptation of the project and thus greater sustainability, were rendered invisible by the rhetoric of objectivity. Martonosi wanted to spark an urgently needed debate on this issue. At the same time, however, he absolutely avoids identifying even a few factors that would delineate those local alternatives. He refuses to reflect on the locally given culture of writing and on local interpretations of procedural objectivity, and he has no interest in what others refer to as "local knowledge." For Martonosi, it must be the Ruritarian project partners who come up with these issues since anything else would simply perpetuate the hegemonic game. Unfortunately, however, the UWEs are the most adamant supporters of the blueprint approach.

I believe, on the contrary, that the technical game is not an instrument of hegemony, but rather the only code available for carrying out transcultural negotiations under postcolonial conditions and the norm of reciprocity. If, however, the results of the technical game are as modest or even as devastating as one could easily be led to believe here, then my position is ultimately even more pessimistic than Martonosi's.

The Politics of Lists

Mlimani, Saturday, September 21, 1997

As I learned from the disillusioned team leader, Mr. T, the situation of the project in late September 1997, fifteen months after it started running, can be summarized in five points:

(1) The consultant is bogged down in the area of organizational improvement because the Ruritanian government has not provided the necessary framework and has not issued any special authorization. (2) The core of the project, the WMS, should have gone into operation in June 1997. At this point in late September it looks like this step will be postponed until December. It is becoming apparent that the old, unreliable data will con-

tinue to be all that is available. The most important aim of the project—to raise the collection efficiency to 90 percent—therefore cannot be achieved. (3) The consulting firm wrote in the midway review that it will terminate the contract if the clients do not satisfy their part of the agreement—that is, supplying the customer data—or propose an alternative solution, and if the repercussions due to the lack of a basic framework are not discussed anew by November 15. (4) The consultants have spent 17.55 person-months (pm) up to the end of the fourth quarter and 22 pm up to the end of the fifth quarter for services that were not provided for in the contract. (5) During the inception phase in March 1996, 20 pm were set aside. The consultant wanted to wait for the results of the study on the technical program in order to utilize the pm as effectively as possible. At the time, the consultant stressed that in October 1996 at the latest a decision would have to be made in order to incorporate these pm into the project planning. In September 1997 this decision is still outstanding. Theoretically there are three options for the continuation of the project:

Option 1: If the agreed-upon aims of the project are to remain unchanged, but they cannot be achieved with the available means, then the *project volume* must be expanded. Initial suspicion has meanwhile been confirmed that this involves comprehensive process control, which must include the technical aspects of the project. If the 20 pm that were initially set aside are used for this, then the project volume of 94 pm or 3 million dollars has been exhausted. This sum must be expanded by the 20 pm that were necessary as noncontractual services. Based on experience up to now, it also appears to be appropriate for the consultant to assume responsibility in the area of customer surveys for conceiving a data correction scheme and for supervising its implementation. The necessary expenditures amount to approximately 20 pm. Option 1 therefore involves expanding the project by 42.55 percent from 94 to 134 pm. Under option 1, the symmetrical reverse case can also be considered: If the project volume is to remain unchanged, the *project aims* need to be reduced to the point that they can be achieved with the available resources. The 20 pm that have already been used for noncontractual services would be offset by the 20 pm that were set aside.

Option 2: If both the expansion of the volume by 42.55 percent and the reduction of the aims are ruled out, *new aims* must be defined and new negotiations must take place to determine *what means* are necessary to achieve them.

Option 3: Theoretically, it is possible to argue that the consulting firm is responsible for financing all the noncontractual person-months it reported.

In this case, there would be claims that the services were either predictable or superfluous. If one also assumes that the consultant is contractually responsible for correcting the customer data, and claims that the delays in the technical program and the lack of a basic legal framework exert no influence on the course of the project, this would lead to the conclusion that the consultant is allotted 94 pm to achieve the agreed-upon goals and is responsible for covering anything over and above that.

Disregarding the question of objectivity, option 1 seems optimal for the consultant and option 3 for the financier and project-executing agencies. In view of this unavoidable initial situation it seems obvious that negotiations would start with claims being made by extreme positions and end somewhere within option 2 with an amicable solution agreed upon by all sides.

That was the team leader's evaluation of the situation. It makes clear that in addition to their epistemic and culture-specific dimensions, lists are also entangled in the power-saturated mechanisms of negotiation processes. In order to perform as well as possible, one needs good lists with hard facts and figures. On the other hand, those players with greater leverage have access to ways and means of asserting their own list as the most "objective" one.

Negotiations : Round One

Mlimani, Monday, September 22, 1997
Ever since Mr. T. became team leader, there has been a so-called UWE meeting once a month in order to get a grip on the project's coordination problems. Today's meeting was a rather unusual one because an NDB delegation is expected tomorrow. As usual the three urban water engineers participated, along with their respective assistant directors. After having already met Joseph Mutahaba in Jamala, I now became acquainted with the engineers Ronald Mapunda of Mlimani and Isaac Mbiti of Baridi. All three had done their basic studies in the engineering department at the University of Baharini, then continued their studies in Madras, India, ultimately receiving their degrees in Europe. The S&P team was represented by T. the team leader, I. the engineer, and Martonosi.

The discussion quickly came to the crux of the matter and the atmosphere was tense: The UWEs claimed that the consultant was contractually responsible for correcting the existing customer data. As regards Mlimani, they said, it was even clear from the very beginning that no customer survey had ever been conducted there, which made it necessary to start from square one. The consultants, on the other hand, claimed that their involvement up to that point in the area of customer data had been entirely noncontractual.

To prove their point, the UWEs referred to the action plan in the inception report. The actions listed in the cases of Baridi and Jamala are: "reorganization of zones, customer survey"; in the case of Mlimani it is: "update of customer master file, reorganization of zones, customer survey." The color of the bars designating the months and duration of the tasks are green and yellow for Baridi and Mlimani. For Jamala the bar is solid yellow. Yellow stands for the executing agencies, green for the consultant. Mutahaba argued that the fact that the bars for Baridi and Mlimani are green and yellow proves that the consulting firm had assumed responsibility for the customer survey. Also, the number of person-months was also listed in the action plan: four fortnights were each drawn in green, which he said led to the conclusion that the consultant planned two months each for the customer surveys in Baridi and Mlimani. From the fact that the bar for Jamala was solid yellow, however, one could conclude that the consultant had assumed no responsibilities there, and Mutahaba was prepared to pay for all noncontractual person-months used for that purpose in Jamala.

Team leader T. responded that the outlays for the activities listed could not be concluded from that table, as he said was obvious from a brief look at the other rows in the table: the sum of the boxes did not add up to the input designated in the catalog of measures. Thus in his opinion the calculation of two person-months was definitely wrong. Besides, he continued, the green boxes referred not to "customer survey" but to "reorganization of zones." With this argument T. brought up a new problem, which triggered Mutahaba's retort: "So why are there no green boxes for Jamala?" T. quickly conceded that the green bar for the reorganization of zones in Jamala had evidently been mistakenly omitted (which had never been noticed before, despite the fact that the inception report had been reviewed three times), and everybody agreed. But the three UWEs would not let go of their conclusion that the green bar for Baridi and Mlimani denoted the consultant's responsibility for the customer survey. They insisted on the validity of their nonwritten *memory*, which in this context seemed rather awkward, since everything had been based on assigning final authority to the written documents and contract.

In attempting to assert his view by referring to an undisputed document, T. had opened up a new can of worms. The catalog of measures that was brought up as a relevant reference was not included in the final version of the inception report. When Martonosi finally found the catalog of measures in the complete version of the original contract, the UWEs contested the authority of the document because the action table of the inception report superseded the old catalog of measures. The team leader was pleased by this move, because it gave him an opportunity to triumphantly read aloud the following passage from the inception report:

The reorganization of billing zones aims to achieve a technical and commercial ideal correlation between town areas (geographic space), pressure zones (technical space), and billing zones (meter reader walking areas and categories of customers in the software package). One of the requirements of this difficult task is a complete customer survey and a fully detailed network analysis. While the customer surveys are underway in three towns, they need to be finalized and the new information needs to be incorporated into the master files. *The customer survey was originally expected to be completed by now without any input by the OIP; necessary additional inputs made by the consultant's staff into this activity now have to be accounted for separately.* The exact amount of this input can be assessed in October 1996.

Mutahaba, the UWE from Jamala, gratefully took up the final sentence, saying, yes, that is exactly how it is. He explained that at the time, the consulting firm had agreed to review the customer data situation in October 1996 to determine what exactly still had to be done and what it would cost. But that never happened. T. countered that the quality of the customer data in October 1996 could not be reviewed because, first of all, the data were not yet entered into the computer—which was the responsibility of the UWEs—and second, the new computers with the necessary software to test for data consistency were not yet operating at the time, which was also not the responsibility of the consultant. When the new computers with the new software finally went into operation in February 1997 and sufficient data had been entered, T. continued, the computer expert S. immediately ran the consistency test in Jamala and the catastrophic findings were reported to Mutahaba. The problem has still not been corrected although the consultant had offered relevant procedural suggestions back in March.

Instead of finding a piece of evidence that everyone accepted, the players of this round moved from one paper to the next. One document refers to the next, and at some point it was obvious to everyone that they were going around in circles. Appeals to the clear evidence of each person's own memory did not offer any escape hatches out of this vicious cycle, either. The participants were only able to agree that the consultant—irrespective of the question of responsibility—should prepare a proposal by mid-October, which would include a preliminary calculation of what inputs by which side were necessary in order to correct the customer master file. The question of responsibility was supposed to be clarified the next day together with the NDB representatives.

With regard to the noncontractual person-months that were invoiced up to June 30, 1997, part of the consultant's demands were accepted after tough debate. The executing agencies agreed that the following person-months had been necessary:

1.90   pm   Installation of the computer networks
1.75   pm   Clarification of the basic legal framework
2.55   pm   Review of the customer data from Jamala

6.20   pm   Sum of accepted person-months

The executing agencies refused to acknowledge another sum of the person-months that had been billed for noncontractual services rendered. In their view, these services were included in the contractual stipulations of the project:

4.90   pm   Implementation of the emergency measures for billing in Mlimani
2.75   pm   Review of the customer data from Baridi
3.70   pm   Review of the customer data from Mlimani

11.35   pm   Sum of rejected person-months

17.55 pm   Total sum of noncontractual pm up to June 30, 1997

Mapunda, the UWE for Mlimani, declared the 4.90 pm for so-called emergency measures for billing to be unfounded because he said the consultant already recognized the problem during the inception phase and therefore could have included it in the regular program. Team leader T. responded that this procedure had been jointly decided between March and September 1996 and he was surprised to hear this interpretation now, a year later, after Mapunda had coauthored the inception report and approved the first quarterly report. The 2.75 and 3.70 pm for reviewing the customer data in Baridi and Mlimani were rejected because they said these tasks were included in the contract as the responsibility of the consultant.

Team leader T. was outraged at the outcome of this debate. During a coffee break he repeated to me—as if repeating it would magically help him fathom the unfathomable—that the question of noncontractual services is part of the quarterly report of any project. One of the most important functions of these reports is to compare the actual and targeted courses of the project, explain any differences that emerge, and find a solution. As in any project, all actions taken were accounted for on a regular basis from the very beginning. The first quarterly report covered the period from July 1 to September 30, 1996, and since that time the person-months for all noncontractual services were listed and individually justified. There were always some differences of opinion but they could always be cleared up. All the points in today's discussions, he said, had been discussed at least once before and a resolution had always been agreed upon. The three UWEs had again confirmed the necessity for 14.25 pm (that had accrued up to March 31, 1997) at a meeting in June 1997, and the financier had already paid for them. For this reason he had felt the only possible aim of today's negotiations could have been to authorize the noncontractual 3.30 pm that came on top of the rest during the fourth quarter (April 1 to June 30, 1997), once the need for it had been explained in the report.

After the final coffee break, discussion started on the consultant's proposal to extend the project a year without increasing the number of person-months. Everyone agreed that the lack of a basic legal framework and the attendant consequences for the project, plus the delay in the WMS, had made it impossible for the necessary one-year trial run under the new conditions to take place in the fiscal year from July 1, 1997, to June 30, 1998. Without such a trial run the clients would not have any opportunity to accurately appraise the services of the consultant. After not being able to agree at all on the meaning of the green and yellow bars in the action plan, everyone readily agreed about the need to extend the project. The meeting ended with a decision to present the list with 6.20 accepted and 11.35 rejected person-months, together with a few other points, to the NDB delegation the next day.

Lists in Power Games

Early in the evening I sat with Martonosi on the terrace of the restaurant, where we went over the negotiations about the person-months over beer and peanuts.

All calculations start with a number, say, 2.75 pm for the review of the customer data in Baridi. Behind this figure lie three others: Mr. T. 0.25; Mr. I. 2.00; and Mr. S. 0.50 pm. Behind these three figures—as implied by the underlying principle—comes a long chain of figures of successively lower levels of aggregation until you finally get down to the hard ground of facts. According to Martonosi, it should be noted that the three statistics in question do not in reality represent countables. First, the actual work represented by the figure—for example, I.: 2.00 pm—is far too heterogeneous and irregular to fit into a standardized system of measurement. The procedure is far too often channeled into different directions than originally planned. It is simply impossible to allocate the daily errands of an expert in a project of this kind to the standard input categories provided. Second, every input by a consultant in a development project is always also triggered by something else: No matter what he does, his input is dependent on the actions of his local partner. For reasons of invoicing it might make sense to orient oneself around the ideal of clear-cut, classifiable actions and clear-cut, distinct responsibilities, but such dividing lines are de facto impossible.

In Martonosi's view, the important thing is not so much the impossibility of these dividing lines, but rather the fact that such divisions are not at all desirable. This is tied to the general phenomenon of *interface definition* in cooperation. In the case of development cooperation, the notorious "hit and run principle" arises from the fact that the consultant maintains the principle of nonoverlapping responsibilities and classifications in order to facilitate billing. If the new systems provided by the consultant are not linked in any way to the local context, they will suffer a rapid demise, because nonoverlapping interfaces were set up when the systems were first developed and installed. Successful links between the new systems and the local context can be established only if overlapping interfaces are used from the outset. If the word "cooperation" is to mean anything at all in the term "development cooperation," then precisely this must be the case. Whereas the "blueprint approach" is at least officially condemned, the accounting logic presupposing nonoverlapping interfaces continues to go unchallenged. The notion that the chain of figures verifying the consultant's input is dependent on work units that have distinct accountabilities and are distinguishable and countable prevents the new systems from becoming anchored in the local context.

If we trace the history of the figure "Mr. I.: 2.00 pm," we see that Mr. I. spent a good deal of this time trying to locate the data—recorded on printed loose-leaf forms—of the previous customer survey, which was conducted in Baridi under the direction of a Normesian engineering consulting company, and then monitoring the data entry into the standard software program Excel. This step was the prerequisite for being able to review the data for consistency and do random sampling to check their validity. This constituted the basis for developing a process for adjusting, supplementing, and correcting the data. After I. had initiated the necessary steps, he passed the responsibility on to the local players, thereby complying with the provisions of the technical game to the letter. After some time it turned out that the operation had petered out because the UWE had recalled the personnel assigned to data entry.

When the IT expert S. came to Baridi in June 1997 to import the data into the WMS, he was informed that only a small portion of the data had been entered. For S., this was a repetition of the situation he had been caught in at the beginning of the year in Jamala. This time, in Baridi, S&P decided to ignore the customer survey and work instead with the old data in order not to lose several more months. Yet, because the old data were of course unreliable and inconsistent, I. and S. developed procedural regulations to correct it using data from the more recent survey and at the same time to review and correct these new data through additional surveys. The waterworks were talked into hiring a so-called customer survey team (comprising an engineer and three unskilled day laborers) for this purpose. The team did not progress all that well with its task, as it constantly produced new errors in the course of the correction process. Project staff members I. and S. therefore started supervising the work. At its own expense, S&P also assigned two people to work fulltime with the customer survey team. Some of the data that the customer survey team arduously corrected were nevertheless already unreliable and inconsistent again by the next round of the organizational process.

The overlapping responsibilities, however, led for the first time to the development of true cooperation. This also revealed something that—from what I had learned thus far—should have been presumed from the very beginning, namely, that behind the repetition of errors loomed a problem whose contours and magnitude went far beyond the approach used up to this point: The waterworks have virtually no formal process control. The staff members do not have an adequate understanding of the principles of systematic, written documentation, formal procedures, or cartography.

At the end of this story of cooperation, S&P's statement of accounts presented a bare figure in a list with other figures: 2.75 pm for verification of customer data in Baridi. In this morning's negotiations the players had acted as if the question were whether the activities represented by this figure lay either left or right of a clear-cut dividing line between the responsibilities of the consultant and those of the executing agencies. In this way the accounting logic tends to camouflage the actual substance of the cooperation: the overlapping areas of responsibility that enable a project to be successful in the first place.

The crux of the problem, according to Martonosi, was that by entering an additional activity category, the need for overlapping interfaces was only further denied. For example, originally there had been three categories to describe work on the WMS: (1) development; (2) implementation; and (3) training. Because the interface with the work performed by the executing agencies was nonfunctioning, a new category was added: (4) verification of customer data. The following actions were included in this new category: (a) analyzing data consistency problems; (b) organizing customer data; and (c) organizing data input. Expanding the action list by a new category acknowledged that an unforeseen development had occurred. But the validity of the list in principle as an enumeration of clear-cut activities to be assigned either solely to the executing agencies or solely to the consultants was thereby confirmed all the more.

Martonosi said that this is the real problem; he called it an *interface trap*. The true reason for engaging in cooperation at all is for two parties to become so intertwined that individual contributions can no longer be identified. But if this is the case, then the work they have performed together cannot be entered into any category of an action list because it cannot be attributed to either one side or the other of a clear-cut interface. The cooperative translation work is thus rendered invisible and unaccountable. This interface trap seems to be caused by something more fundamental than the accounting logic of projects. The communication between the consulting firm and the executing agencies is based on the assumption that both parties work within a given framework that they share and accept without question, according to which there can only be a few corrections here and there within the technical game. I argued—and this time Martonosi did not object, even though it is a touchy issue for him—that the list autophagy syndrome cast doubt on whether this common ground really exists. This kind of doubt, however, is taboo in the trading zones of the project, as well as in anthropology. This taboo in turn causes precisely the main point to be factored out of negotiations, that is, list autophagy. Factoring out list autophagy ultimately means failing to address the key issue of how some notion of objectivity can be established as standardization through fixed procedures in a context in which such a notion is highly implausible. Instead of dealing with this central question, the actors of the trading zone agree on a trivialized notion of objectivity as the correspondence between representation and reality—as if everything could be reduced to questions such as "Do we have 1,000 or 3,000 water meters?" Martonosi calls this kind of blind spot that emerges in the technical game the *objectivity trap*. If this is correct, then the interface trap is primarily a consequence of the objectivity trap.

One would think that of the three parties—financier, executing agencies, and consultant—the executing agencies would have the greatest stake in preventing the project from failing, in particular from failing because a model that might work in the West has been transplanted into their context without any reflection. This is because they would be the ones primarily held responsible for not engaging in that reflection. Whereas all commentators of the organizational field and all officials emphasize that model transfer is the primary cause of failure, we have seen here that it was precisely the Ruritanian executing agencies who advocated model transfer. They wanted at all costs to assert the technical game with its accompanying blueprint and clear-cut interfaces. By insisting on the clear-cut interfaces in the technical game, the UWEs delegitimized the consultant's efforts to achieve something behind the facade of the technical game. For the consultants, this meant that ultimately they would end up in an *accountability trap* vis-à-vis the financier. The consultant would be unable to invoice some of the services that he actually had to provide in order to keep the project going. This means that at some point the consultant would stop performing these services, so the main problem—list autophagy—would then no longer be tackled even unofficially.

Martonosi drew a little diagram in his notebook depicting the vicious circle identified here (see figure 5.3). The players in the field—as far as they believe in the technical game—are convinced that it is possible to get out of the vicious cycle by presenting irrefutable facts and figures. In particular Shilling seems to believe that the number chain can be traced from the statement of accounts all the way down to the firm ground of facts. Once that is done, the controversy about accountability is settled through objective means and things can move forward. The best opportunity for this act of objectification is, according to Shilling, in the midway review, because it is possible at that time to discuss things openly that initially had to be omitted. Martonosi, in contrast, is convinced that the controversy can only be sensibly brought to a conclusion through an act of self-reflection, if the list of input categories is expanded to include the following line: "How does this action list function?" Only after the answer to this has been agreed upon can any agreement be reached about everything else. He is the only one in the project who supports this idea.

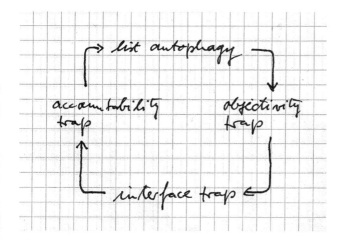

Figure 5.3
The vicious circle (from Martonosi's field notebook).

At the end of the day in Mlimani, however, the following thought went through my head: Every player knew this morning that the controversy over the noncontractual person-months was not really about clearing up a past situation. Everyone must have been aware that discussing this controversy was actually preparing the ground for a decision about how the project can continue. The prescribed choreography was clear: the consultant gathers arguments to prove that the only reasonable decision would be to follow option 1 (increase the input and retain the aims, or conversely, retain the input and reduce the aims). The executing agencies argue that it would be impossible to choose anything but option 3 (retain the aims without increasing the input). In principle, both sides must know that in the end the decision will have to be somewhere within the scope of option 2 (modifying both input and aims). Of course, there are various possibilities for compromises within this option, with very different impacts on the budget. In search of one of these compromise formulas, the UWEs evidently overbid their hands, as Mr. S. put it afterward.

Is it even possible to follow Martonosi's suggestion and discuss openly how the lists function within the negotiation process, given the power-saturated conditions in which such a discussion would inevitably occur? How could players then still be able to make any advantageous moves? And how can negotiations ever come to a close if the arguments employed do not claim to be absolutely objective?

Negotiations : Round Two

Mlimani, Tuesday, September 23, 1997

The next morning the same group reconvened at 8 o'clock to meet the NDB delegation, which was made up of Mr. P., the project manager, and Mr. W., the hydraulic engineer at the development bank. So now all three parties in the special configuration I had heard so much about had come together.

Without any discussion Mr. P., head of the NDB delegation, led the session. There was evidently no prearranged agenda. First Mr. P. reported on the current state of affairs. Then it was noted that the consultant, together with the UWEs, were supposed to present a use proposal for the 20 pm that had been set aside. Between the lines (that is, according to the U script) this actually meant: The executing agencies who received the money for the 20 pm for consulting services from the NDB evidently still did not know how best to use these person-months, so that here the consultant had to step in. This subject led right to the two sore points: the noncontractual person-months that the consultant had billed, and the responsibility for the correction of the customer data.

The agreement (according to the O script) that had been made a day earlier between the executing agencies and the consultant about accepting 6.20 pm was acknowledged by the representatives of the financier (according to the U script) as a submission for critical review. Mr. P. considered all noncontractual person-months to be questionable and requiring extensive justification. Although on the surface it looked like this intervention was aimed at the consultant, it questioned the abilities of the UWEs who had placed their signatures under the 6.20 pm the previous evening. The UWEs, however, preferred simply to say nothing, engaging in a kind of "active passivity."[8]

Negotiations on the noncontractual person-months in the area of the customer survey necessarily led to the more general question of areas of responsibility and the overlap between these areas. That finally brought the meeting to the main issue. In the midway review of August 15, the consulting firm wrote that it would cancel the contract by November 15, that is, after the contractually stipulated ninety days, if the three executing agencies could not provide adequate answers pertaining to their failure to perform their contractual obligations, particularly in the area of customer data. Because Mr. P. had been told in advance of the purpose of the meeting, the S&P team assumed that he would give his interpretation of the contractual situation in order to assist the two other parties out of their month-long deadlock. When Mr. P. instead admitted that he was not yet familiar with the contractual situation and would first have to peruse the files in detail back in Urbania and possibly confer with the legal department, he left the team speechless. It was decided that by October 15 at the latest, the consultant was to present three papers to the executing agencies: (1) a detailed justification of all noncontractual person-months; (2) a proposal for the use of the 20 pm that had been set aside; and (3) a proposal for correcting the customer data.

As the participants were already packing their bags, Martonosi realized that it made no sense for S&P to even begin working on the three papers without knowing in advance how the financier views the contractual situation. If the financier were in the end to say that the consultant was responsible for the customer surveys in Baridi and Mlimani according to the valid contract, S&P would cancel the contract as they planned by November 15 without any further ado. Mr. P. felt this argumentation made sense. Mutahaba, the UWE of Jamala, was going to be in Urbania the following week on unrelated business. P. agreed to go through the contract before meeting Mutahaba and report on the findings to him. Today's meeting was already adjourned at 11 o'clock because the NDB delegation had allotted only three hours for it before they had to fly to the next project in Ruritania. While the UWEs and the two gentlemen from the NDB embraced each other as they said goodbye, the two Normesian parties nodded somewhat awkwardly to each other without even shaking hands.

Whereas at yesterday's meeting with the UWEs great importance had been attached to writing down and having all responsible parties sign all decisions, today the spoken word apparently sufficed, and afterward, so did the unwritten memory of what had been agreed upon orally. Later, however, I found out that the assistant director of the waterworks of Mlimani had kept minutes of today's meeting. He would give his draft of the minutes to T., the team leader, in a few days for any linguistic corrections and after that it would be circulated and filed.

After dinner, I drove with Martonosi to the Grand Hotel, where we drank beer on the dilapidated terrace, sat through frequent power outages, and listened to oldies. After the first glass S., the IT expert, and Mbiti, the UWE of Baridi, joined us. We talked about how the two meetings went and cracked acerbic jokes about moves that had backfired. It seemed to me that Mbiti was trying to emphasize that the harshness of the dispute should not at all affect the confidence he has in the consultant. What Mbiti was most concerned about, however, was the fact that Martonosi had openly contradicted the two gentlemen from the NDB and even pointed to possible errors they had made. From his perspective, Martonosi had violated a standard of politeness, and he was trying to prevent him from making such faux pas in the future. The IT expert, S., a man from Costa Rica who had studied in the United States, countered this time with a cultural interpretation. He thought the conduct of the NDB delegation was an example of what he called typical Normesian authoritarianism. This kind of behavior could thrive in Ruritania, he said, because Ruritanians are subservient to influential people. What was grotesque about the whole show, he continued, was that the NDB managers were taken in by this kind of behavior and interpreted the Ruritanians' politeness as agreement. Later in the evening T., the team leader, also arrived at the Grand Hotel for a beer. He felt that S&P had not been able to score any points at either meeting because the matter had been poorly prepared. Most of his moves hadn't panned out because the documentation he referred to was either incorrect or incomplete, which he could not have guessed since he was not responsible for them at the time. I was preoccupied with a message implicit in the morning's proceedings. It sounded the whole time as if the project were being carried out solely here on the ground and solely by the consulting firm. The UWEs and the NDB did not seem to ascribe any influence on the course of events either to themselves or to each other. They seemed actually convinced that this was all about a service they were paying for.

Baridi, Tuesday, October 14, 1997

Shilling returned on October 10. Martonosi gave him a rundown on the latest negotiations and used the opportunity to reiterate his theory on the main problem facing the project. The two meetings on September 22 and 23, he said, had simply confirmed that lists with facts and figures are not the point of departure of negotiations, but their result. Shilling, on the other hand, felt that the time had now come to present hard facts and figures. Since his return, he had spent countless hours at his notebook in the project office in a trailer on the grounds of the Baridi waterworks. He was trying to prepare a few understandable tables as an initial summary of what had been achieved so far and of the resources that had been used, thereby producing the first of the three promised papers.

Martonosi was working on the other two papers, one a proposal for the use of the 20 pm that had been set aside and another for correcting the customer data. To this end he was following the work of the customer survey team in Baridi and trying to draw some conclusions from the error analysis.

A comprehensive process control system was supposed to be developed using the 20 pm. One of its tasks in the course of its routine operations would be to review and correct the customer data at regular intervals. This approach stands in contrast to the solution that, as a result of the pressure of the circumstances, was looming in the project, namely, a new, one-time collection of customer data in the form of a "large-scale raid" of the entire urban area under the direction of the consultant. Martonosi was sure that this solution would only appear to eliminate the problem. Even if a new and complete customer survey were able to achieve a relatively high recall ratio, most of the corrected data would soon be corrupted again. But most of all, a new survey would shift attention away from the real cause of the problem—namely, list autophagy—to a trivial point within the technical game. The message conveyed by this shift would clearly be: "Everything ran smoothly once the consultant corrected and entered the primary data." That would put us back at the same point where the entire drama had started almost a quarter of a century ago, when a well-equipped technical facility was set up in the countryside in Jamala and no one bothered to integrate it into an organizational process.

The organization could learn something from the experience, Martonosi felt, only if the correction of the customer data were integrated into the everyday work routine with printed forms and cards, and if this integration were given high priority in the project. This is because the quality of the data is dependent only on the consistency with which the organization members follow standardized procedures. They would gradually realize this if they chose this approach. Of course things would proceed more slowly and deviations from the plan would be more serious. But if it is common knowledge that concentrating on spectacular increases in productivity—in this case "90 percent collection efficiency"—is one of the main causes for project failure, then it should be possible to avoid these errors. There are already fitting slogans for the better alternative in the jargon of development cooperation: "capacity building" or "institution building." And that is precisely what this project had started out to do in the very beginning.

Work on the three papers, which were supposed to be completed by October 15, proceeded feverishly, despite the fact that all efforts to obtain prior clarification of the contractual situation had been unsuccessful. S&P evidently could not afford simply to refuse to prepare the concept papers and thus risked losing the two additional person-months that went into this task. Since NDB's response, which was supposed to be sent along with Mutahaba, the UWE of Jamala, around October 1, was not forthcoming,

Martonosi called the bank in Urbania on October 8. He took heart when he heard that according to the NDB interpretation, the UWEs were responsible for providing the data. They also agreed to extend the ninety-day deadline by two weeks, until November 30, 1997, so the October 15 deadline for submission of the three papers was abandoned by implication. When on October 14 Mutahaba came to Baridi for a meeting, however, all hopes were again dashed. Mutahaba said that in Urbania they believed that correcting the problematic data situation was part of the contract from the outset. A few days later a fax finally arrived from Urbania. It stated that the UWEs were responsible for carrying out the data collection. But regarding the past and future input by the consultant in the area of data improvement, they were still waiting for the consultant's comprehensive report.

Baridi, Saturday, October 18, 1997

Yesterday, the day before we were leaving, Martonosi called Mr. K. of the Prime Minister's Office in Baharini, whom we had visited in his office a few weeks ago on September 4. K. reported proudly that as of November 1, 1997, the wages of the employees of the waterworks of Baridi, Mlimani, and Jamala would no longer be paid from the civil service payrolls via the regional governments. The Ministry of Finance had issued a special regulation according to which the wages for a transitional period would instead come directly from the Ministry of Water. Apparently, K. made an intelligent move here. Now he could say to the World Bank that the wage lists of the civil service had been reduced by the number of employees of the three waterworks. But insofar as the wage amount would be provisionally paid through the water ministry, nothing had really changed. The Ministry of Finance, however, would not be able to maintain these special transfers to the water ministry in the long run. Pressure would thus be placed on the water ministry to complete the final step, making it more likely that wages would soon be financed through revenue from water sales. In any case, the water ministry had now lost its comfortable position of being able to make decisions regarding the municipal waterworks (most recently by appointing the board members) while someone else pays the bills. This move finagled by K. was also smart in that he cautiously addressed the financial autonomy of the waterworks in order to avoid the danger of insolvency. And of course Martonosi was also personally relieved since ultimately he would have shared the responsibility for that insolvency.

That afternoon Martonosi said goodbye to Mbiti, the UWE of Baridi. The two of them have known each other since 1992, when Mbiti was still working in a lower position. It seems to me that they are both unhappy that they cannot develop a closer friendship. Mbiti expects Martonosi to acknowledge him as a peer, an expert and "professional" in a global business. Martonosi, on the other hand, expects to be acknowledged by Mbiti as someone who understands African perspectives and experiences. But as long as this understanding includes the recognition of something other than the qualities of a professional, namely, the recognition of an elementary cultural difference, Mbiti cannot fully satisfy Martonosi's expectations. In contrast to other fields, cultural difference is not a readily acknowledged category in the global epistemic community of development experts, but rather one that is almost always bracketed out. Conversely, Martonosi cannot fully satisfy Mbiti's expectations because he would also have to abandon one of his most dearly held convictions, namely, that the universal professionalism aspired to by Mbiti is undermining the future of Africa.

In the evening we said goodbye to Julius Shilling. Here as well there was an air of melancholy. Because the semester was about to begin, Martonosi had to return to the university. Shilling had to stay on because this work is his life and not just his research. The next day the chauffeur brought us to the airport. This time the airbus that took us to Urbania was named "Marilyn Monroe."

Negotiations : Round Three

Urbania, Thursday, November 27, 1997
A good month after returning from Ruritania, I was curious enough about how things were going that I arranged an appointment with Martonosi. We took a long walk in the city park as he caught me up on the news from the project. On November 22 the executing agencies and the consultant in Mlimani had finally finished writing up the draft on the continuation of the project.

Shilling had worked out the proposal together with the UWEs based on the assumptions that, first, the inputs could not be changed (so the maximum of 94 pm remained unaltered) and, second, the already used, noncontractual person-months were to be financed from the 20 pm that had been set aside in March 1996. Third, the proposal also assumed that correction of the customer data was not part of the contract thus far, so 9 of the 17 pm that were still available in December 1997 would be taken from other tasks in order to use them to solve the data problem. Starting on December 1 the consultant was going to organize a new survey of customer data, first in

Baridi and then in the two other regional capitals. As soon as this prerequisite has been fulfilled, the consultant will fully install the WMS, activate the deactivated security procedures, and put it into operation. Team leader T. and computer expert S. are available for this task. Everything else will be put on hold and the experts will be withdrawn. Of particular importance is the fact that the proposal contains a relatively precise list of the tasks assigned to the UWEs. The UWEs assumed a considerable share of the work to prepare the list, which led to a joint interpretation of the problem. The joint proposal emphasizes that reducing the project to the successful installation of the WMS is an emergency measure that could easily become a farce if the project never goes any further. A suggestion was made to implement the rest of the original plan with new resources in a second phase, assuming that the legal framework has been set by that time. The proposal was signed in that form by both parties as an addendum to the project contract.

Today, November 27, the reaction of the NDB to the addendum prepared in Mlimani arrived. The financier considered the proposal to be correct in principle, but the consultant was invited to Urbania on December 3 to clear up several remaining questions. The ninety-day deadline (for the consultant to withdraw from the contract) following submission of the midway review on August 15, 1997, which had already been extended once, was therefore extended by another two weeks until December 15.

Urbania, Wednesday, December 3, 1997

Martonosi and I picked up Shilling from the airport early in the morning to drive together to the Normesian Development Bank. The meeting took place in a windowless room in the Countries Division; except for a short lunch break, it lasted the whole day. The development bank was represented by five gentlemen. P. was the project manager in the Countries Division; W. represented the Technology Division—I knew both of them from Mlimani. O. was the representative of the Sectoral Division, who had been involved in the Mlimani workshop in 1996. New to the group were U., the section head and direct superior to P., and U.'s boss, Division Head A., who was present for part of the meeting. Section Head U. facilitated the meeting, and there was no agenda available in advance, just like in Mlimani. In front of me were six, and sometimes seven, men sitting around a table, negotiating something that could not be brought to the room. "The project," that is, the reality being discussed, thus had to be represented by proxy. This occurs in two ways: via written papers and tables, and via nonwritten memory and narrative.

Figure 5.4
Allocation of person-months up to June 30, 1997.

At first it looked as though the entire negotiation were centered around the lists with facts and figures that Shilling presented, that is, the appendices to the addendum that the executing agents, the UWEs, had signed on November 22 in Mlimani as a conception for the continuation of the project. All of the negotiating partners were given numerous pages with copies of the tables. And all of them had stern expressions on their faces as they hunched over the pages that were the representation of the project in this room (see figure 5.4 for an example). It soon became apparent that Shilling's remarks referred to a different version of the tables than the ones he just had distributed. The conversation kept getting bogged down in futile attempts to interpret the figures correctly. Shilling, who was the main author of the tables, said little during the discussion, compelling the other participants to work up to his level of understanding in order to interpret the tables. Little by little people started to show their displeasure.

Participants repeatedly complained that it was hard to understand the copies they received for the simple reason that the original tables had been printed in three colors, so it was virtually impossible to make heads or tails out of the black-and-white copies. One participant said it made no sense to discuss something in December using figures from June. Another one remarked that the new format of the table of individual actions might be a bit clearer than the format that was used in the third and fourth quarterly reports, but that overall, the new presentation format made it more difficult to make comparisons, so no developments could be recognized.

Project manager P. was the member of the group who otherwise suffered most directly from the lack of transparency in the tables. One of his most important responsibilities was to work through the tables at regular intervals to uncover any possible errors or deliberate chicanery. His boss, U., relied on P. for this because it would be impossible for him to review the figures of all the many projects he supervises on his own. P. kept a low profile at this meeting and seemed to enjoy that his colleagues finally saw how hard his regular job was, working with Shilling's tables. You could tell from the looks of the participants that some of them had a creeping suspicion they were being blinded with manipulated figures. The escalation in the offing was subtly prevented by discussion facilitator U., who steered the players away from fundamental suspicions of deception back to the actual substance that the figures are supposed to represent.

The difficult debate about numbers unfortunately did not yield any conclusive figures. No one added up the interim results to achieve an overall picture, let alone a sum of the total number of person-months needed and accepted, and the corresponding amount of money involved. Instead, the consultant was given the job of amending the draft addendum that had been prepared in November in Mlimani, incorporating the results from this meeting. This vague result was even less binding when one considers that Section Head U. did not have the last word. During the nerve-racking work with the tables, a series of telling dissonances arose within the NDB team.

The engineer W., for example, claimed that the consulting firm had obligated itself in the last version of the inception report (of September 1996) to accomplish the declared project objectives using 74 pm. With that he was saying that the often-mentioned 20 pm had not been set aside in the inception phase, but instead had been canceled. This was the only point in the entire discussion at which Shilling raised his voice, rejecting the interpretation as false. The discussion facilitator ignored his Technology Division colleague's interpretation of the contract without a word. Of course there were differences of opinion as soon as discussion moved on to correcting the customer data. P. and W. held the position that the consultant was violating the contract by now suggesting a reallocation of person-months from other areas for a new customer survey. W. lay the familiar letter of May 1996 from the NDB on the table in front of him, pointedly tapping one particular section with his index finger, and read aloud: "S&P formulates the requirements . . . for the customer survey . . . and presents a suitable proposal for the additional input at no extra cost."

This, W. claimed, proved that the work the consultant had been causing such a furor about since August 1997, and for which he now wanted an additional 9 pm, was in fact part of his original contract and included in the 74 pm; and, he added, the consultant never responded to this comment from May 1996.

At this point Martonosi took the floor. He said that there were two ways to interpret this sentence. Either the phrase "at no extra cost" is regarded as the immutable point of departure and the meaning of the sentence as a whole is based on that. Accordingly the sentence could only mean that the consultant has to inform the executing agencies what data are needed to localize a customer. To the extent that this could be accomplished easily, it could be expected that the consultant would do it "at no extra cost." Or the immutable point of departure in determining the meaning of the statement is interpreted to be the clause "S&P will formulate the requirements for the customer survey." That would consequently mean that the "requirements for the customer survey" include a conceptualization both for the collection of customer data and the integration of the new data with the old data, and with the process control for the entire system. But if "requirements for the customer survey" refers to this whole package of elaborate and logistically complex interventions, the extra phrase "at no extra cost" makes no sense at all, especially when the point is not merely to deliver a recipe but specifically to supervise the implementation. According to the most recent calculation, conceptualization and supervision of the implementation amounts to 9 pm. Neither one of the two ways of interpreting the sentence made any sense since the two parts are mutually exclusive.

The sentence is nonsensical for yet another reason, Martonosi said impatiently. In March 1996 the executing agencies had insisted that their data had already been collected and were accurate. The development bank expressly supported the position of the UWEs at the time. Therefore, the consultant had no reason to expect that two months later, in May 1996, the NDB would present a totally opposite opinion and speak of "additional input." How could "additional input" be necessary for something that has already been completed? And why should this "additional input" be rendered "at no extra cost," if that is the main point of the whole project, namely, to correct the inadequate process control in the waterworks? Instead of hitting the NDB with these inconsistencies, the consultant reviewed the situation between May and September 1996 and entered the findings in the inception report. There one could read a clear response to the NDB letter of mid-May 1996: "necessary additional inputs made by the consultant's staff into this activity now have to be accounted for separately. The exact amount of this input can be assessed in October 1996." The inception report had been accepted by all parties and is valid until further notice. So how can W. now refer back to a letter of May 1996?

Martonosi's forceful words violated one of the rules of the game. Instead of explaining and justifying his actions as consultant in an objective, businesslike manner, he shifted the focus to the actions of the financier. However, these were not at issue here. Instead of winning advocates for his argument, Martonosi earned a warning from Division Head A., who had joined the group. On the other hand, W. did not gain any supporters for his position either. Discussion facilitator U. once again inconspicuously but cleverly maneuvered the negotiations further as if neither W. nor Martonosi had ever spoken. The 9 pm that the consultant was demanding for conceptualizing and supervising the collection of new data and integrating them into the system were no longer questioned, although this was never expressly stated, nor did P. and W. ever explicitly abandon their contrasting positions.

Negotiations circled largely around the figures that Shilling had provided. No one had any other figures and no one doubted his calculation method; at most, more transparent tables were desired. Nevertheless, there were doubts about the reliability of the figures. The skepticism arose due to images and narratives that were not included in the actual discussion. This became particularly obvious when Division Head A. joined the group for an hour. One of the things he wanted to know was: "How far along are you with your customer survey, Mr. Shilling?"

Shilling responded in a way he felt appropriate for the negotiations: They now had a good plan, and implementation had been well underway in Baridi since November 1. As of December 1, a crash program for the collection of new data had even been started, he added. He didn't mention a word about the background to this new development and did not make any reference to what had already been said. The staff members working for A., that is, project manager P. and section head U., remained silent. Everyone seemed to agree that their boss, A., did not need to know what was actually going on here. Martonosi, on the other hand, interpreted the appearance of A. as an indication that the management smelled serious trouble and wanted to make its own picture of the matter. Instead of the routine procedure of being informed through the official channels of the hierarchy, A. was willing to hear personally what Shilling had to say. But Shilling had already finished giving his report.

Martonosi again couldn't stand the silence. He tried to explain how things were really going. He started with the inception report of September 1996, mentioned the midway review of August 1997, finally coming to the consultant's contractual insecurity, which has been protracted since that time. With a vehemence that seemed out of place in this down-to-earth, business setting, he pilloried the NDB's refusal to disclose its interpretation of the contract, thus touching on the incompatibility between the positions that had just been presented: On the one hand W. was trying to win points with his letter of May 1996, and on the other hand U. was not questioning the necessity for additional person-months to conduct the customer survey. The division head interrupted Martonosi's animated intervention with the attitude of a person who had already mastered far greater problems. Mustering all the authority that the honorable institution granted him, he declared that such details should not be discussed while he was there, since his time was short. With that his visit had almost come to an end. After saying goodbye and turning to go, he glanced at Shilling and remarked, with a slight irony in his voice: "I am counting on the ability of so many highly paid experts to manage to collect some customer data; okay, Mr. Shilling?"

Urbania, Thursday, December 4, 1997

After yesterday's meeting I accompanied Martonosi to Mr. von Moltke. As we walked down the corridors of the Countries Division, Martonosi exchanged a few words with an employee he knew, who proudly mentioned how well the process of privatizing the railroad in Uganda was going. Someone else reported about success in a residential housing project in Windhoek, and in the elevator he ran into the friendly railroad engineer that he had worked with in 1990, the first time he was in Ruritania for the NDB. Von Moltke had gone into early retirement a few days ago and had already cleared out the office of the director of the sub-Saharan Africa Main Division, where I had spoken with him in July. After a few introductory remarks, he asked about the present state of affairs of the project, although he seemed to be already informed. It turned out that he had joined A. for lunch after A. left our meeting. Von Moltke, A.'s former boss, apparently understood that the executing agencies had gradually become dissatisfied and nervous because S&P kept demanding more and more person-months, although there was hardly anything to show so far. Especially as regards the customer survey, no end was in sight. Martonosi was searching desperately for a fitting response. If von Moltke got his information from the "apparatus," it was impossible for an outsider to question the validity of the statement. After all, he personally spent years helping to build up that apparatus and, until a few days ago, was its director.

The answer nevertheless turned out to be point-blank: "The project is in danger of failing. This is because, in the supplementary negotiations, the executing agencies are overtaxing the structurally weak position of the consultant and the NDB is not mediating this time." Von Moltke looked surprised; he cautiously reminded Martonosi that his interests lay elsewhere. The NDB, he continued, had brought him together with Shilling to help in the process of overcoming the centralized bureaucracy in the Ruritanian water sector. He said the project was on the right track toward introducing market economy solutions and has offered considerable momentum. Common, trivial details such as a delayed customer survey or a few disputed person-months should not be of concern to Martonosi. We continued to talk a while about general issues of development cooperation and also of retirement, and the conversation continued on a very friendly, almost personal, level.

That evening, after Shilling's plane had departed for Mercatoria, Martonosi invited me to dinner. We tried to recap the course of the negotiations. Martonosi found it enlightening that the NDB staff was composed of people who had realized the dream of a position in civil service. Although the development bank was not directly part of the civil service, as a state bank that pays an annual wage supplement of two months' wages, it is a desirable employer that offers high job security. Like most members of the civil service, the younger NDB staff, in particular, looks down with a feeling of ethical superiority at those who simply "make money," such as consultants. During a break in the negotiations, someone had said that the development bank was accountable to the taxpayers for "every dollar," so it was essential that the consultants really only receive the share they truly deserve. From this perspective, the development bank is always on the side of the common good, whereas the contractor is on the side of maximizing private advantage. That is the diametric opposite of the idea of privatization, the spread of which the NDB promotes throughout the world. According to the logic that makes a governmental agency a natural ally of the taxpayer and by the same token makes the contractor the taxpayer's natural enemy, it must be assumed with respect to the Ruritanian waterworks that the old, state-run, bureaucratic organization was in fact the best one.

In terms of the project, we had reached the final point in the vicious cycle during the meeting. From list autophagy, or rather its invisibility within the technical game, we progressed via the objectivity and interface traps to the accountability trap, in which the project was now stuck. This vicious cycle was only a special case of a general phenomenon. Of course it is easier to refer more precisely and with greater reliability to aims, means, costs, and procedures in the second half of any project than in the first. In most cases a certain openness and lack of clarity in the beginning should not be seen only as a deficit. Instead, an initial uncertainty is necessary to develop the optimism needed to get the project going. All parties that want a project but still need to win over additional alliance partners are well advised not to illuminate the entire background down to the last detail. Such efforts generally bring contingencies to light that can undermine faith in the feasibility of the project.

This results in a common pattern for projects: In the beginning one leaves open as much as possible, and in the end one concludes as much as absolutely necessary for accounting purposes. About midway through the project, plans and contracts must be adapted to actual developments so that it is possible to conclude what needs to be concluded. Projects routinely suffer from a midlife crisis at this point, because the uncertainty that is desired at the start prevents a simple comparison of the actual state achieved with the contractually agreed-upon target state. It then becomes inevitable that all parameters have to be redefined: the given situation, the targeted solution, the contract conditions, and the procedure for assessing the achieved state. This redefinition, however, must at the same time be denied, for otherwise it would give the impression that the project is unpredictable and, consequently, financially incalculable. That would result in the loss of an indispensable prerequisite for being able to carry out future projects at all.

This aporia can be resolved by rhetorically appealing to contractual stipulations and facts on an official level, and secretly redefining the entire situation on an unofficial one. This rhetoric gives the impression that the subject of dispute in supplemental negotiations is whether there are five or three apples in the basket when in reality negotiations are trying to ascertain what is meant by "basket" and "apple," who is responsible for putting "apples" into the "basket," and how the outlays should actually be accounted for. Without these basic redefinitions that have to be cleverly concealed, it would hardly be possible to survive the midlife crisis. From this perspective, the facilitator of today's negotiations was a real virtuoso of this technique.

In this case, things proceeded de facto the way they do with any compromise. One side claims they can only satisfy the agreed-upon aims if the resources are increased, and the other side disputes this. In the end they meet halfway. At first 94 pm were allocated, but from S&P's perspective, 134 pm would be necessary. If the individual points that were discussed are added up—which was not done during the negotiations—the total amounts to roughly 122 pm. The NDB therefore bargained the consultant down from 42.55 percent additional pm to 29.78 percent. In exchange, the bank accepted more modest project objectives. This compromise was based on a redefinition of the interface between software design and data collection. The NDB representatives, however, now acted as if nothing had changed in the basic approach. In order to preserve the impression of predictability and calculability, it had to look as if the original contract continued to cover everything. That is why, to the very end, all controversies over the correct interpretation of the contract remained unresolved.

Because the midlife crisis of a project can only be overcome by redefining the entire context, written documents play an ambiguous role. In part they must be "filed away" in order to create leeway to negotiate unforeseen and deviating solutions. This move, once again, cannot take place in the open because the fiction of sticking to the original procedure based on clear-cut documentation needs to be maintained. At this point, oral knowledge comes to bat, which remains rhetorically concealed, however, because this form of knowledge plays no official role in the technical game. If a weaker player nevertheless questions the argumentation of a stronger player as being based on oral knowledge, a warning will be issued. This is how a strikingly simple sentence—that is false in an extremely complicated manner—managed to remain undisputed at the end of the negotiations with the top of the NDB hierarchy: "The executing agencies are dissatisfied with the consultant because he has still not corrected the customer data."

Urbania, Sunday, April 5, 1998

At the meeting on December 3, 1997, Shilling agreed to write a new contract addendum with the conclusions they had come to and the solutions they had proposed. By the end of December he had still received no response to the document he had punctually submitted and was thus forced to terminate the project contract as of January 31, 1998, because otherwise he would have forfeited the security deposit (the 10 percent of the paid consultant fees retained by the financier). On January 21, 1998, there was another crisis meeting in Urbania, at which it was again resolved that Shilling would submit yet another, slightly altered proposal. When Shilling again received no immediate answer from the development bank, his own commercial bank ceased to believe that S&P was still in business at all. With immediate effect it refused to advance any more payments for the company. On February 6, 1998, the S&P consulting firm had to suspend all business activities pending further notice.

Of all the people in the project, Shilling was the one who believed most steadfastly in the unfaltering strength of facts and figures. In his final invoice to the financier, he submitted charges for his most recent services and the security deposit amounting to a total of $120,000, a sum that was based on figures that have already been reviewed several times. The financier rejected these charges and submitted its own invoice, according to which the consultant owed an outstanding $30,000. In this offsetting invoice, all of the noncontractual person-months that Shilling had withdrawn from the negotiations for tactical reasons were now classified as non-allowable. The consultant, who meanwhile was threatened with insolvency, was unable to counter this demand. Because the development bank asserted outstanding claims from the consultant, the bank guarantee of over $72,000 that Shilling had deposited could also not be paid out. On April 4, 1998, Shilling filed bankruptcy for the S&P Corporation before the Mercatoria Local Court. Interpretation of the facts and figures is currently in the hands of an official receiver. According to Normesian bankruptcy laws, this receiver is legally charged to recover especially those outstanding debts that can be collected at a reasonable expense and that have good prospects of being recovered.

Urbania, Sunday, October 11, 1998

S., the former S&P systems analyst, who was in the meantime back home in Costa Rica working freelance, made an offer to the Ruritanian executing agencies to install the now corrected customer data using the software he had developed for the waterworks. The executing agencies rejected this offer. Instead, another consulting firm was commissioned to determine the objective facts and figures of the current state of affairs, because no one really knew the status of the project anymore. To find out how things continued, I would have to embark on new investigations. My story, however, ends here.[9]

TRYING AGAIN

6 Metacode—Cultural Code

Preliminary Remarks

In the prologue, I introduced the protagonist Edward B. Drotlevski, a fictive anthropologist engaged in a research project. Drotlevski then proceeded to introduce us to three more main characters: Johannes von Moltke, who spoke for the financier in chapter 1; Julius C. Shilling, who spoke for the consultant in chapter 2; and Samuel A. Martonosi, who recounted his experiences as project anthropologist in chapter 3. In chapters 4 and 5, Drotlevski offered his own observations directly from the field in the form of edited excerpts taken from his field diary. In this concluding chapter, I will review the various reports and summarize their results. In addressing the different issues, I concentrate on representational practices, technologies of inscription, models of objectivity, and code switching.

Three of the characters in the text have been presented as a chain of observers. Shilling observed the realities to which the project had to be adapted, Martonosi observed Shilling's realism, while Drotlevski observed Martonosi's relativist deconstructions. In bringing together their observations here in this final chapter, I will address the relationship between objectivism, relativism, and constructivism in negotiation processes. Martonosi's observations have supported the view that the code of objectivity was a mask disguising the hegemony of the "donors." This concealed hegemony was seen as ultimately responsible for the fact that local perspectives never played a significant role and that the project was therefore doomed to failure. Nevertheless, Drotlevski's reports suggest that these negotiation processes can function only under the premise that players observe this code of objectivism. In closing I will examine this contradiction in greater detail.

A Performance with Two Scripts and the Role of the Consultant

The countries receiving development aid are officially considered sovereign nations that determine their own fate. Official discourse declares that the project-executing agencies and the people directly affected by the project know best what they need and how they can help themselves. They are the most competent experts in their own affairs. Development experts must first learn from these people before they are able in turn to teach them anything. The current catchwords are self-determination, participation, and local knowledge. From this vantage point, development cooperation consists in providing people with the resources they lack: loans, technology, and expertise. According to this principle, the desired results are inaccessible to those who allocate these resources because only the sovereign recipients possess the capacity to achieve such results.

On the other hand, the fundamental premise of development cooperation is of course that Western industrialized democracies not only have the highest standard of living, but also the most advanced knowledge. They can and must make this knowledge available, along with favorable loans and technology, to the poorer nations of the world so that the latter can finally experience sustainable development. If one were to assume that the relevant knowledge is valid only within the sociocultural and political frame of reference of the US–European world but is misplaced in Africa, then the fundamental premise of development cooperation would be rendered invalid. If development cooperation means anything at all, it must ultimately involve organizing a transfer of knowledge and resources. This knowledge includes first and foremost knowledge about the laws of social development everywhere and always. According to this principle, results of this kind of intervention are accessible to those who intervene.

As the reports presented here have demonstrated, the art of development cooperation consists in skillfully avoiding the irresolvable contradiction between the accessible and inaccessible sides of this work. In order to do this, development cooperation is performed with two different scripts. According to the guidelines of the official script (our O script), this involves transferring (along with capital and technology) technical knowledge, which is socially and culturally neutral and which anyone can acquire through appropriate training (How can I re-coil and repair an electric motor? How do I use Excel spreadsheets?). In contrast, according to the guidelines of the unofficial script (our U script), development cooperation involves the transfer of precisely the kind of knowledge that is supposed to transform the basic forms of social order (What can be regulated by the market economy? How is political legitimacy constituted? How should loyalty conflicts between kin and the common good be resolved? What is procedural objectivity?). In performing this play, it is possible to switch from one script to the other, depending on the particular situation. There is yet another reason that speaks for this arrangement, and it is located on the level of the practical realization of projects.

By definition, a national development bank works in business fields where normal commercial banks do not. This means that their borrowers are selected because of their credit-*un*worthiness. According to the *progress narrative*, the aim here is to set increased productivity in motion by providing loans, technology, and expertise, which will lead to the credit-worthiness of that organization in the future. Once this is achieved, the development bank will have made itself superfluous and in this way have achieved its own objective. However, credit-unworthiness as the prerequisite of an economic enterprise means at the same time that borrowers are not really trusted. If the recipients were actually able to deal with this input in a responsible manner, they would not be in this predicament in the first place and would not require the input. From this vantage point, the reasonable solution would be to place the loan recipient under the temporary supervision of the financier. This alternative is impossible, however, as a result of the postcolonial *emancipation narrative*. This is not a tragedy for an institutionalized development bank, but actually a kind of blessing. Assuming direct responsibility for the implementation of projects would inevitably damage the image of this infallible professional institution. To resolve this dilemma between external monitoring and the self-determination of the project-executing agency, the development bank introduces the figure of the consultant. It is here that the U script assumes a concrete function. According to the U script, the financier is the actual but secret overseer of the consultant. To a great extent the financier determines autonomously what the consultant should do for the project-executing agency.

To do justice to the difficulty of this performance with two scripts, a further dilemma must be considered. For reasons mentioned earlier, the party that plays the role of the donor is predestined to attach conditionalities to this transfer, indeed is compelled to do so because it must account for its own actions back home. However, against the backdrop of the emancipation narrative and the O script derived from it, the setting of conditionalities is not really tenable. For this reason, the financier and the project-executing agency easily and elegantly reach agreement on procedures that would allow the conditionalities to appear as if they were simply the more rational strategy. (The workshop in Mlimani in October 1996 was analyzed as an example of this and contrasted with later developments in the project.) Although more direct language has been employed since the early 1980s (with the introduction of so-called structural adjustment and policy-based lending) and has become even more pronounced since the end of the cold war, everyone avoids explicitly articulating the basic rule that all of the players recognize: "You will get something from me if you do what I tell you to." The figure of the consultant is also necessary because, whereas financiers have to adhere to the diplomatic protocol and to their own long-term goals, consultants are contracted to assume responsibility for doggedly sneaking in conditionalities for the allocation of resources. (The example of this drawn from our case study was replacing the Revolving Fund Act with a privatization provision that had no legal basis in Ruritania.)

It is clear that a performance using two different scripts enables and encourages an infinite series of tactical script changes. Those players with a situational or a structural advantage are often able to switch scripts with minimal risk, thereby creating further advantages for themselves. (Drotlevski's reports were full of such examples; ultimately the dynamics of the project's development were determined by script-switching. This began with the NDB's letter in mid-May 1996 and ended with the script change between the meeting in Mlimani on November 22, 1997, and the meeting in Urbania on December 3, 1997.) The possibility of tactical script-switching ultimately results in habitualized mistrust.

This mistrust caused by switching scripts is intensified by two additional factors. In transferring responsibilities from the state-owned development bank to the private consulting firm, the suspicion inexorably arises that the more knowledgeable consulting firm might in its role as agent systematically deceive the development bank in its role as principal and in this way create advantages for itself. Conversely, the consulting firm observes how the cumbersome bureaucracy of the bank botches up individual projects through stubborn compliance to its own guidelines, thereby robbing the consultant as well of success and profit. (In our case, the controversies that unfolded around the midway review and the noncontractual person-months are eloquent examples of this mistrust.)

Finally, this lack of trust is also tied to the relationship between the consultant and the project-executing agency. As a result of the errors and shortcomings of the executing agency, the consulting firm is able to expand its responsibilities and in this way maximize its own advantage. In light of this circumstance, the executing agency tends to disguise its own shortcomings and regard the consultant's error diagnoses as exaggerations. Because the project-executing agency has access to better local knowledge, it can rather easily set the consultant on the wrong track. Failure to address the weaknesses of the executing agencies not only undermines the project, but can also ruin the consulting firm's reputation, since its own success is judged according to the accomplishments of the project-executing agency. Just as the project-executing agency has to presume that the consultant magnifies the problems, the consultant in turn inevitably fears that the project-executing agency plays them down or conceals them. (The never-ending drama about customer data in our project provides a telling example of this structurally determined problem.)

If projects are not to be reduced to an absurdity by this game and yet at the same time are not supposed to consist of a powerful player imposing something on a weaker player, and if projects also cannot be limited to a financier simply transferring money to a recipient, then the only remaining option is the negotiation of an at least provisional consensus. Given the two scripts and the habitualized mistrust, however, the triad of financier–project-executing agency–consultant proves particularly unsuited for this in structural terms. As we have seen, the probing question "Where are they going with this move?" holds the restless players permanently on tenterhooks. And when, in spite of it all, the negotiated definitions of reality, solutions to problems, and interfaces of areas of responsibility occasionally last for more than just one move in the game, this seems rather miraculous under the given circumstances—that is, if we look for the answer solely in the realm of trust and consensus. The established framework that every game requires for players to predict the next move is instead produced in a different way—through *technologies of trust.*

The Metacode and the Technical Game

The reports have shown that the global organizational field of development cooperation operates according to a *technical game* oriented around the central dichotomy of effective–ineffective. The technical game is part of the narrative of progress. Progress is perceived as the ability to design social development through the ever-increasing capacity to control the natural and social worlds by constantly improving science and technology. Even if this progress narrative has already been tarnished for some time now— and has been identified by some as the cause of the two great totalitarian systems of the twentieth century—it has retained its self-evident validity in development cooperation. This is all the more remarkable given the fact that development cooperation has without doubt provided the most powerful evidence that social development does not follow predictable rules and hence cannot be established according to a plan. Figure 6.1 outlines the inevitability of the technical game once the world is conceived according to the unquestioned premises of the progress narrative.[1]

A negotiation according to the rules of the technical game can begin in any of the four squares of figure 6.1 and develops as an iterative process that moves back and forth through the squares. In square one of the chart (bottom right) negotiators are confronted with the problem of not knowing exactly what they are aiming at, what they can accomplish, who the concerned and the responsible parties are, who can legitimately speak in their name, what the other negotiating parties want, and finally how they can arrive at a consensus about these questions. They are even confronted with the problem of not knowing exactly how they can acquire the necessary knowledge to answer these questions. If the negotiations are to have any chance of succeeding, players must move through square two into square three and from there into square four, where they then have reached their goal.

In square two, players engage in research to gain knowledge about the situation, the relevant actors and stakeholders, the existing problems, and the possible solutions. This is facilitated by choosing the best procedures

Knowledge

		Certain		Uncertain	
Consensus	Complete	Problem: Technical Solution: Calculation	4	Problem: Information Solution: Research	2
	Contested	Problem: Disagreement Solution: Negotiation/Coercion	3	Problem: Knowledge Solution: move to Field 2	1

Figure 6.1
Knowledge improvement in technical games.

and instruments that are available worldwide. (To this end the NDB commissioned a social-scientific analysis of the Ruritanian water sector in the summer of 1992 and a feasibility study on the organizational development of the waterworks in Baridi, Mlimani, and Jamala in the fall of 1994.) The increase in scientific knowledge acquired in square two does not, however, imply evaluative and prescriptive propositions dictating specific actions. In square three, therefore, the negotiating parties need to reach an agreement about their preferred goals and how these are best to be achieved based on the acquired knowledge. Our reports demonstrate how "negotiation" and "coercion" can easily overlap here. (This square includes, for example, the negotiations between the NDB and the water ministry as well as other ministerial authorities in preparation for the project.) Once a prescriptive agreement has been reached, the only remaining step is to calculate the technical aspects of the cooperation. If players manage to reach square four, they are in a position to engage in rounds of negotiations on purely technical aspects that can be calculated and that trigger concrete decisions and actions. The first rule of square four states that if difficulties arise, players should try to solve them within square four; if this does not work, returning to square three is acceptable, but retreating to square two or one should be avoided at all costs since this would normally mean terminating the project. (The midlife crisis of our project in the fall of 1997 occurred when players in square four wanted to resolve problems that would have required returning to square two, which was officially impossible on the basis of the accounting logic and the power configuration among the players.)

The four-field scheme in figure 6.1 is constructed on two axes: differentiation of knowledge along the horizontal axis and consensus along the vertical one. This distinction corresponds to the difference between facts and values, or between what is true and what is good. In more abstract terms, the key difference here is between a denotative language game and an evaluative or prescriptive language game. The distinction between the two, however, is a heuristic one, and the four-field scheme offers an apt illustration of how each square of the iterative process is actually an intersection between knowledge (fact) and consensus (value). The work in square two and square four does not consist of purely technical activities occurring beyond normative considerations. Like all technical activities and enabling devices, they are inextricably interwoven with values and interpretations that depend on a frame of reference that includes epistemic, normative, and material dimensions. The converse is also true: disagreements on goals and means in squares one and three cannot be reduced to evaluative and rescriptive factors, since these depend fundamentally on technical-scientific positions.

The iterative process depicted in figure 6.1 is generated by the rules of the technical game. Our interlocutors Martonosi and Drotlevski agree that the technical game exerts a hegemonic side effect or is itself an instrument of hegemony. Both agree as well that this hegemonic effect tends to obstruct the desired cooperation and thus to undermine the sense of the entire enterprise. However, they differ markedly in their interpretation of the causes of this hegemony and its relation to objectivity. The technical game presents itself officially as based on the naive realism of everyday thinking. Accordingly, it is possible for objectively true statements to circulate unproblematically between all of the different frames of reference because the validity of these statements is grounded in external reality and is consequently independent of any and all frames of reference. This means, however, that universally valid statements must be formulated in a universally valid language, which Martonosi calls a *metacode*. First Shilling and later Drotlevski presented a series of cogent arguments about the indispensability of the technical game with its assumption of realism and its metacode. Martonosi, on the other hand, argued repeatedly that the insurmountable problems of the project were a direct consequence of the technical game.

For example, the basic phenomenon of list authophagy had to be suppressed because it cannot even be conceived within the technical game. The admission of list autophagy would ultimately result in admitting other taxonomies and forms of referentiality, that is, admitting cultural differences, which in turn would undermine the conditions of the technical game with its claims to universality. Without the premises of the technical game, players would never get beyond square one. Conversely, it is also true that the project's profound ineffectiveness resulted precisely from this denial of list autophagy or, to formulate this positively, from the denial of local forms of knowledge. Martonosi was also able to demonstrate (most recently on the basis of the final negotiations at the NDB on December 3, 1997) that, contrary to the rhetoric of the technical game, the "facts and figures" employed to resolve the controversies—in squares three and four—were not prerequisites worked out in square two but rather the results of conclusions that had been drawn in the squares three and four by other means. In order to examine more closely the relationship between hegemony and the technical game, between coercion and negotiation, and between power and objectivity, a closer look at the centers of calculation as the point of origin of the technical game is necessary.

The Metacode as the Language of Centers of Calculation

In chapter 3, we learned from Martonosi that centers of calculation produce new knowledge in the sense that that they bring together things that in reality only exist separately and place them in a common context. In this way, new connections are discovered that would not have been thought of at any of the separate locations. The production of this kind of knowledge is dependent on being able to collect accurate representations of distant events, things, and people at the center of calculation. To this end, three basic conditions have to be met. As our reports have demonstrated, the representations must be mobile, immutable, and combinable.

Since it is generally impossible to physically move the objects in question to the center of calculation, a transportable representation (a proxy or a depiction) is required. In the reports presented in the previous chapters, the problem of *mobility* was encountered in different variations. In each of these cases, however, players in location x had to negotiate and reach a decision about the reality in a distant location y. It is not possible to hold monthly meetings with the almost ten thousand water customers in Baridi in order to influence their payment practices. Nor is it possible to check the water meters individually every morning or even to transport them to the materials warehouse at the waterworks every three months in order to maintain an overview of their condition and whereabouts in Jamala. This issue also came up during the meeting in Urbania on December 3, 1997. Because it was impossible to present "the project" by physically bringing it to the meeting, the participants had to make do with several sheets of paper. Confined to a windowless room in the center of calculation, they brooded for an entire day over these pieces of paper, which were supposed to represent the events that had taken place in Ruritania from the beginning of June 1996 to the end of June 1997.

In addition, a mobile representation of something has to arrive unaltered at the center of calculation. In our project, every player who wanted to get a picture of a distant reality was plagued by the suspicion that the available representations had been falsified somewhere along the way. Some players had problems locating their customers, others in documenting their person-months, and still others in tracking the success rate of their projects on the other side of the globe. Immutability is a problem insofar as mobile representations wander as a rule through many hands and many contexts and are inevitably altered thereby before making it to the center.

In order for a center of calculation to achieve the desired increase in knowledge—so that more is seen at the center than could be seen at each of the individual locations on its own—these mobile and immutable representations, which arrive from all over and at different times, must be brought together in a final step. The necessary *combinability* can be achieved by using taxonomies, criteria for selection and ordering, and standardized procedures for measuring and aggregating, which are prescribed by the center as part of a metacode with universal validity. However, everyone involved must comply with the metacode introduced by the center and the corresponding forms of reporting from the very first steps of the representation process for the simple reason that both the mobility and the immutability of representations can be guaranteed only through strict standardization. The center must ensure that it does not receive a myriad of incompatible reports, as these can hardly be made compatible after the fact. In the first place, the center of calculation simply could not handle the workload; second, the error rate would be far too high because the external referents are located thousands of miles away. (*Notes and Queries on Anthropology*, published in 1874, is an early standardizing work of this kind.)

Insofar as a development bank is supposed to channel the money it administers into appropriate projects and is thus forced to make prognoses, and insofar as it has to account for the correct, effective, and efficient use of this money, it must be a center of calculation. Its main concern is the production of *translocal knowledge* according to the aforementioned three basic principles; this knowledge enables the bank to monitor and control from a great distance, which in turn allows it to account for its own actions. This is not only officially expected of a development bank; even a critique of development cooperation will necessarily raise the same demands and will itself have to comply with principles of verification and accountability. No one would seriously suggest that taxpayers' money should be distributed for international development or for the scientific evaluation of development work without some form of accountability.

The reports cited from our project initially landed on the desks of the responsible project managers in the NDB Countries Division. Here they were collected with other reports from all the various projects occurring within that country. As a result they already appeared in a different light than when they were viewed independent of one another. Contours became apparent that were not visible from the perspective of the individual projects. The reports then proceeded to the desks of the section heads, where they were collected with the reports from three or four other countries. This shift in context brought still other aspects to light. The information from larger regions was combined and analyzed at the desks of division heads, before they were passed on to the head of the sub-Saharan Africa Main Division (previously Mr. von Moltke), who combined and aggregated the information such that he was able to issue the following statement: "Since the introduction of the structural adjustment program things have been progressing in sub-Saharan Africa" (this is "proposition A" mentioned in the prologue to this book). In the Sectoral Division, the reports, in contrast, were combined according to sectors, which meant that all reports on water systems landed on one desk, while all reports on organizational improvement programs landed on a second desk, and all reports on energy projects on a third desk, and so on. In the Technology Division, engineers developed their own particular insights by bringing together projects from around the entire globe that employ comparable technological solutions.

On their journey through the divisions and hierarchy of the centers of calculation, these reports became, on the one hand, increasingly thin, which is the purpose of this procedure. The information was transformed according to particular rules into a new form and was thus *reduced*. On the other hand, through this same procedure the information was also *amplified*. Although project managers in the centers of calculation received only individual project reports about local events, in the end comprehensive statements could be issued. This amplification process presupposed that the specific comprehensiveness of individual project reports was subjected to a progressive reduction in order to paint an overall picture.

The work of a development bank as a center of calculation thus consists in perfecting a knowledge technology of this kind to the greatest possible degree. The materials for this task are the submitted reports, which means that the heart of a development bank is its archive. All the reports are collected there and must in principle be compatible. To the extent that this combinatorics works, the center of calculation has a considerable head start over the local knowledge contexts of individual projects. (The tension between the consultant and the financier after the Mlimani meeting of September 23, 1997, resulted from the conflict between the necessarily divergent requirements of local and translocal knowledge in assigning person-months to scheduled and contractual or noncontractual activities. This tension, however, only erupted into a dispute because the actors incorrectly believed that they were dealing with the absolute distinction of "true" and "false" as defined by correspondence theory.)

As an institutionalized organization, a development bank has an interest in concealing the fragility of its own knowledge technology. The ritualization of its performance evaluation as well as its undisputed role as an obligatory passage point and the accompanying definitional power facilitate the concealment of this fragility. However, another form of concealment—one that does not require much effort on the part of the center of calculation—is decisive here. The technical game settles this issue automatically by officially invoking the objectivity of correspondence theory and thereby denying its own fragile representational practices. Whenever any dissatisfaction with the representations arises, the rules of the technical game simply recommend returning to the reality in question in order to compare it with the disputed representations—as if it were possible this time to forgo the entire laborious procedures and simply let reality speak for itself.

This recommendation is based on the premise that we have twofold access to external reality, once as it simply exists and once as it appears in our (procedurally conditioned) representations. What occurred, however, between a particular past event in Baridi, Mlimani, or Jamala and the representation of this event in Urbania—the introduction of a metacode, reduction, and amplification—is pushed into the background, and our undivided attention is directed to the correspondence between the initial elementary representations and the external referents. This gives the impression that the challenge is simply to avoid making mistakes when anchoring our representations in reality: Was it three or five gallons/customers/water meters/leaks/unpaid bills/person-months/dollars, and so on? The actual work of the development bank as a center of calculation—namely making information transferable without deformation for the purpose of control from a distance by introducing a metacode in the center's archive—is concealed through the public performance of a technical game. This not only deflects any critique of the pitfalls and shortcomings of the mediation work; it also renders invisible all of the hegemonic effects on the local frames of reference.

This can be seen more clearly if we recall that the purpose of the information transfer to the center of calculation is to guide and oversee processes (of development) from a distance. Using the machine metaphoric of the technical game, this dimension of oversight can be described as follows: The center of calculation is connected to the project through a drive chain, which as a cogwheel sets in motion the next drive chain located within the project-executing agency (thereby activating development). The project-executing agency should in turn function as the thrust for further

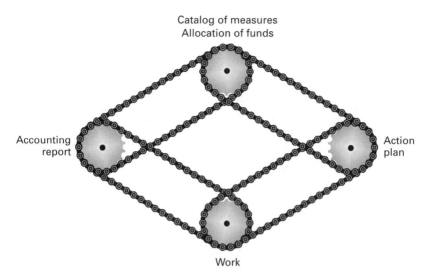

Catalog of measures
Allocation of funds

Accounting
report

Action
plan

Work

Figure 6.2
First drive chain in the technical game.

developments that come to encompass an entire society so that the project can ultimately continue in a sustained way without any external drive mechanism.

The first drive chain (figure 6.2) proceeds as follows: In the center of calculation, money and the catalog of measures agreed to contractually are available as a means of control. In the project itself, the catalog of measures is translated into an action plan, which is in turn materialized as actual work. Through this process a quarterly accounting report is submitted to the financier, who then decides if further monies will be made available or not, depending on the project's success.

The official purpose of this drive chain is to activate a second drive chain, which operates within the organization of the project-executing agency. In the case of our organizational improvement program (OIP), this initially meant installing a nonexistent drive chain (or repairing a disrupted one). The central problem of the Ruritanian waterworks was that it was not economically viable because it could not maintain its systems of control from a distance. The representations contained in the files of the waterworks repeatedly degenerated into a farce, so that the waterworks were able to collect only a fraction of their water bills because of a lack of information.

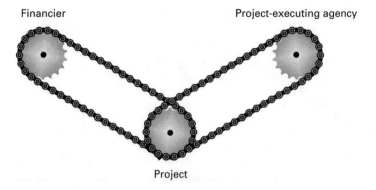

Financier   Project-executing agency

Project

Figure 6.3
Second drive chain in the technical game.

For the first drive chain to generate and activate the second drive chain, however, it is not enough to simply have the consultant work to achieve this. He has to work together with the project-executing agency. The activities of these two parties should be coordinated like that of two cogwheels: The movements of the former are transferred to the latter in a one-to-one ratio, so that the drive of the first chain (figure 6.2) is directly transferred to the second chain. Figure 6.3 provides a visualization of this process using the machine metaphoric of the technical game.

The conceptualization of the project as a technical game, as a transfer of motion in a drive mechanism through chains and cogwheels, is a hegemonic process insofar as the locally self-determined definition of how one wishes to live and work does not appear at all in this model. The reports drawn within the scope of our OIP, however, have shed light in particular on two aspects that call into question the hegemony thesis. Our interlocutors were able to confirm that the accounting logic of the project did in fact proceed exactly according to figure 6.3. This basic pattern can be recognized most clearly in the fact that the indicators of success used to evaluate the consultant's work were oriented around accomplishments that are supposed to be carried out not by the consultant but by the project-executing agency. Consultant and project-executing agency are thus supposed to be linked like cogwheels. Yet whenever the O script applied, it was always emphasized that self-determination and participation are the first priorities for the sake of better project results. The imposition of a preexistent model, which would serve the interests of the center of calculation but might infringe on the local context of the project-executing agency, is explicitly rejected by all statements within the O script. This is the diametrical opposite of linking two cogwheels together. In fact, as our interlocutors have also confirmed, the project-executing agency was indeed able *de jure* to determine the course of the project in a sovereign manner and even *de facto* exert the most decisive impact on the course of affairs—albeit in a very different way than officially conceived. Seen in this light, it is astonishing that hegemony could be established in this game at all. Presumably a certain consent on the part of those being patronized played a role here.

Furthermore, our reports indicate that the machine metaphoric of the technical game disguised something absolutely essential (and for this reason will be the focus of the next section): We appear to be dealing here not so much with drive chains whose links are held together firmly and rarely torn apart, but with *translation chains* whose segments are difficult to hold together. From this perspective, the metacode proves to be an expression of the efforts to hold the parts together. If this aspect is taken into consideration as well, it is possible to analyze more precisely the Ruritanian executing agents' consent to the dominance of the technical game and its metacode.

The Metacode as the Language of Translation Chains

We can use the image of the data highway to depict another dimension of the official representation of the knowledge available to a center of calculation. In our case, the data highway runs between the project—that is, the Ruritanian waterworks—and the desk of the project manager at the NDB (figure 6.4).

The implicit idea is that reality can be adequately represented through a simple bipolar schema: here the reality (of the project in Ruritania), there the adequate representation (in the center of calculation in Normland). In order for information such as that in figure 6.4 to flow from A to B and back again without any deformation, several things have to occur, as we have seen repeatedly in the reports taken from the organizations and the trading zones of our project. Although the most important task for the center of calculation and the other participating parties is to facilitate this flow of information, the work required to do this is kept behind the scenes. The official representation gives the impression that one could simply pack up a load of data from the waterworks in Mlimani into twenty-five file folders and drop them on the project manager's desk at the NDB. The project manager, however, would insist that the real translation work still has to

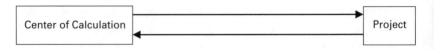

Figure 6.4
The ideology of information flow.

be done: The content of the twenty-five folders has to be transferred onto a one-page table, which must be both objectively correct and compatible with his office's frame of reference. In other words, it must be composed in the prescribed metacode. Figure 6.4 depicts the official version of the process, whereas figure 6.5 presents the behind-the-scenes work required to maintain this surface.[2]

For information to flow within translation chains, elementary facts have to be classified, measured, translated into numbers, aggregated, and combined. The question mark on the right side of figure 6.5 represents one instance of this process, which in our case connects the waterworks (context C with the broken line) with its environment. This connection is represented by a question mark because it is necessarily fragile, as we will see in a moment. Also, the connection is never completed because every elementary representation can in principle be divided into smaller elementary particles. In this sense, the "first" representation of a thing or an event is always re-represented as it moves along the chain (from right to left). Consequently, a repeated *translation* or *transformation* of data takes place, which is the condition for its arrival at the center in an immutable and compatible form that can be understood and evaluated as information. In our case, the process that occurs between waterworks and environment (at the right end of the diagram) is repeated once between waterworks and consultant (between context C and B [the solid line]) and then again between consultant and financier (between context B and A [the dotted line]). In principle, the chain does not end here but continues on to the left, concretely for instance in the connection between the financier and the responsible ministry and from there to the political apparatus, so that it is impossible to identify any fixed endpoint.

However, the contexts A, B, and C in the diagram also represent the successive steps of information processing within one and the same organization, for instance, the aforementioned steps between the hierarchy of desks at the development bank. The central aim of our OIP was to intervene in how the translation chain (A, B, C) functioned within the

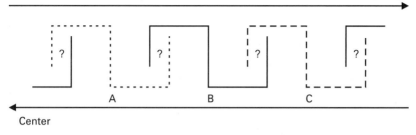

Center

Figure 6.5
The translation chain.

waterworks. The so-called second reality of the waterworks, that is, the representations circulating in management-level offices (context A), was supposed to be made more reliable in order to ensure the economic viability of the organizations. The issue here is identical to that of the twenty-five file folders in the previous example that the project manager at the NDB did not want to receive. The directors of the three waterworks were confronted with precisely this problem: If they wanted to make a decision (we have heard about this repeatedly in detail), they had (in context A) no available facts and figures (about context C) that were clearly arranged, manageable, and also correct. For every decision, even about something absolutely mundane, they had to study all of the files, or even worse, commission completely new studies about the state of affairs in their own enterprise. They lacked a procedure for producing the key figures that are required whenever issues have to be dealt with, monitored, and decided from a distance.

The quality of these *key figures* does not depend first and foremost on the correspondence of the elementary primary data with external referents (at the right end of diagram 6.5). They must of course correspond, but there is nothing problematic or noteworthy about this. It is only after this point that the actual work begins—that is, transforming elementary data into meaningful information. Key figures depend by definition on their stated purpose and on the procedure selected *a priori* (in context A), and not on their correspondence to reality. If, for example, it was reported that in 1995 the waterworks of Mlimani had a system loss of 70 percent (in other words, the revenue that was brought in covered only 30 percent of the water produced), nobody would dream of verifying this figure by looking for a single external referent outside of the office. Or if a consultant invoiced 17.55

noncontractual person-months, there is also no possibility of confronting any naked reality with the figure of 17.55. Finally, if the NDB claims 75 percent of its projects are successful, a trip visiting all of the projects would not be a valid review procedure.[3]

For the waterworks as well as for the project and the development bank, the real key is rather that basic primary data must be prepared in various ways such that these data make sense and fulfill their function on the desks where the decisions are ultimately made (in context A). The quality of key figures is demonstrated by the fact that they reduce localized, complex realities (from context C). The more universally valid they are supposed to be, the less they take into account the diversity, complexity, materiality, and particularity of a localized reality. Instead, the amplitude they attain is all the greater at the end of the chain (in context A) than at the beginning. In all of these cases, when one reflects on the objectivity of a statement or a number, the point is evidently the following: It must be possible to retrace the precise individual steps that were taken in the course of the original process. In doing so, however, it is not a correspondence between reality and representation that is confirmed, but instead an *adherence to procedural rules* during the re-representation in the individual sections of the chain and especially between the sections (where there are question marks in the diagram). The correct procedure never results directly from reflections and reviews carried out in terms of correspondence theory. The reverse instead is the case: Only through the procedures and the individual steps of re-representation can we review the results. Between these two points lies the indispensable *mediation work.*

Mediation work consists of an indissoluble mixture of methodological and taxonomical preliminary decisions (without which one would be lost in a chaos of unordered reality), paradigmatic or theoretical provisions (without which it would be impossible to distinguish plausibilities), the instrumental-rational weighing of interests (without which even the simplest decisions could not be made), and power-based exertion of influence, value references, and basic ideological decisions (which can never be completely excluded). The farther away one moves from individual local cases (for instance, the work of the project's computer expert in March 1997 in Jamala) in the chain of translation by means of reduction, and the closer one thus moves to amplification at a remote center of calculation (for instance, the meeting in Urbania of December 3, 1997), the more re-representations rely on the sum of established knowledge with all of its provisions.

As soon as the role played by this procedurally directed mediation work is recognized, the following—I reiterate—becomes clear: The facts and figures have to be transformed (reduction/amplification) in order to be transferred. At every step of this transformation process, however, there were also different options than the ones that were actually selected. This is because, contrary to the official discourse, the successive re-representation steps are not based entirely on the elementary facts and figures. Instead, this re-representation is always also determined by the general framework of the respective links in the chain (A, B, or C). Particularly in the transitions, in passing through trading zones located between contexts, these re-representations are influenced by the conditions of the negotiation processes. The question marks in figure 6.5 represent this moment of inevitable indeterminacy.

The reports presented in this study contain numerous examples of negotiation processes in which players sought to replace these question marks with "hard facts" that would remain valid for all participants over the long term in order to keep the chain from breaking. However, the validity of a conclusion worked out laboriously in one trading zone was repeatedly called into question again in the next trading zone. The most conspicuous example of this occurred in the two successive meetings in Mlimani: The conclusions reached during a first round (between C and B) on September 22, 1997, were transformed back into question marks in a second round (between B and A) on September 23, merely because different people participated in the second round. The same thing happened again between November 22 and December 3, 1997. The reports suggest the following thesis: The main reason why the chain continues to break should be sought where true cooperation occurs between the project-executing agency and the consultant and where interfaces between the respective responsibilities are negotiated.

The Metacode as the Language of the Trading Zones

Development cooperation occurs in a global arena in which players seek to cooperate under conditions of *heterogeneity*. As the reports have illustrated in detail, however, heterogeneity exists in terms of not only interests but also basic orientational knowledge, which on a preconscious level contains provisions about semantics, plausibility, evidence, causality, relevance, legitimacy, and ethics. Nevertheless, in order to cooperate successfully under these conditions, a method must be found that allows reliable transitions from one frame of reference of orientational knowledge to another. In other words, an epistemic link has to be found for all frames of reference.

Within the network of those organizations engaging in development cooperation, some organizations prove to be centers of calculation. To the extent that money flows from these centers into the network and this flow has to be accounted for, these organizations must develop modes of monitoring and controlling from a distance, and this in turn presupposes a metacode that claims to be valid in all frames of reference. As soon as the metacode is established as an epistemic link and the rules of remote sensing, monitoring, and control are considered valid, the global strategy referred to in this study as the technical game begins. The *hegemonic subordination* of local contexts, which are incorporated into the translocal context of development cooperation by means of the technical game, occurs primarily through the practice of disregarding the implicit orientational knowledge of these local contexts.

One of our protagonists, Martonosi, argued that the *power* of the metacode comes from centers of calculation that have successfully occupied obligatory passage points. Or one could simply say that the power of the metacode lies in the *money* that the centers of calculation allocate, insofar as they tie this allocation to the use of their metacode. Another protagonist, Drotlevski, arrived at the following conjecture in the course of his field research: The aforementioned interpretive schema diverts attention away from the actual function of the metacode in two regards. First, interpreting the metacode as a hegemonic disciplining instrument gives the impression that development cooperation is actually possible without a metacode of some kind. Second, this interpretation never questions whether it is necessary that the metacode of the technical game brackets out the implicit orientational knowledge of local contexts to such an extent that cooperation inevitably fails.

The reports of our project are replete with evidence that the project-executing agencies do not in fact feel oppressed by the metacode of the technical game. It turned out, on the contrary, that the technical game provided the only basis for cooperation under the conditions of heterogeneity. In order for Ruritanian experts to be able to negotiate definitions of reality with US–European experts on an equal footing, to make joint decisions, and to act collectively, they have to have access to a metacode. Seen in this light, the metacode is not only a necessary prerequisite for the business of centers of calculation, but also a condition for the indispensable reciprocity of players in a trading zone. This is an irresolvable ambiguity that simply has to be accepted and endured.

The archetypal situation of practical development cooperation is that of negotiating concrete definitions of problems and solutions. These negotiations result as a rule in decisions that are supposed to be implemented. For this process to occur at all, a firm ground of facts, a binding standard, and a metacode have to be assumed. To borrow from Alfred Schütz, the negotiating partners must find a "natural attitude" that allows them to believe that the objects of the external world are in principle identical, and have the same meaning, for all interlocutors. Negotiating partners must, according to Schütz's thesis of the *reciprocity of perspectives*, at least presume that the other partners experience the world under the same conditions as they experience it themselves. According to this premise, there can be normative and aesthetic differences between players' frames of reference, but not fundamental differences in their implicit orientational knowledge.[4]

The tricky combination of setting conditions and offering assistance, of hegemony and equality, and of the superiority of one party over the other results in a perfidious form of camouflaged disciplining. The technical game is the means of this disciplining. Nevertheless, under the given circumstances, the technical game appears at the same time to provide the sole effective means of defense. Only according to the premises of the technical game does it appear legitimate to accept a preexistent model at the gates of one's own empire in order to maintain the integrity of that empire. The messenger may leave behind some technical instructions and perhaps even brief the recipients on how to use the model, but then he is supposed to go back to where he came from without posing any further questions. The appropriation of the model subsequently succeeds as an internal and sovereignly self-determined affair.

The fundamental character of this process can be best understood through an analogy. The everyday practice of development cooperation occurs in contexts that Drotlevski called *trading zones*. The archetype of the trading zone is the marketplace, where goods and services are traded. For a successful exchange to occur, it is essential that all aspects that could result in disruptive complications be disregarded. The ideal of the market implies minimizing relevant factors and information to what is absolutely necessary (or formulated conversely, maximizing externalities that should be omitted from exchanges). The extreme case of this minimalizing strategy can be seen in what anthropologists call *silent trade*. One party deposits its goods (for instance, salt) at a particular location and then departs; the other party comes to the location and leaves as much of its own goods (for instance, corn) as it considers commensurable near the goods of the first party and then departs as well. If the first party agrees to the trade, it picks up the goods left by the second party. Or, if it is not satisfied, it leaves them there and the second party has the option of adding something. If this works, the exchange can be repeated in the future; otherwise it is broken off. Modern markets of monetary societies work in principle in the same way: The parties exchanging try to consider only information that concerns the exchange. As a rule this means those factors that can be articulated in money.

The same mechanisms function in trading zones concerned not with bartering goods but with cooperation under conditions of heterogeneity. For cooperation to be able to take place, all elements and information that could interfere must be factored out. This requires first and foremost—analogous to economic markets—a *pidgin trade language* that lets the information that is absolutely necessary be communicated and all superfluous information disappear as much as possible. In the trading zones of development cooperation, the technical game with its metacode has been established as such a trade language. In each particular case, from silent trade to cooperation regarding the metacode, the question arises whether this minimizing strategy has not perhaps gone too far, so that the trade or the cooperation appears successful only at first glance. Over the course of time it could easily prove that relevant information was inadvertently excluded, resulting in unpleasant surprises.[5]

In development cooperation, errors in minimization are as a rule discovered during the course of the project. Because of the accounting logic, correcting these errors requires enormous effort and frequently fails. With reference to the rules of the technical game and the unquestioned validity of the metacode, it was simply presumed in our project that all the participating parties had a firm footing in the basic cultural technologies of writing and bureaucratic documentation. The submitted reports, however, have demonstrated that not all the parties had mastered these basic technologies, which resulted in so-called *list autophagy*. Because recognition of this syndrome would have violated the reciprocity of the players and thus threatened the technical game as dictated by the diplomatic protocol of the arena, the issue was treated as taboo or translated into a problem that could ostensibly be resolved within the framework of the technical game through the transfer of advanced technology and expertise.

When list autophagy became taboo, it resulted in an objectivity trap between the consulting firm and the project-executing agency. The two parties repeatedly agreed to certain procedures (for example, regarding the correction of customer data), which then proved to be ineffective a short time later. The main reason for this was that the implicit basic assumptions about procedural objectivity did not exist on the Ruritanian side. However, for reasons of reciprocity and attributability, this difference had to be denied. Consequently, the objectivity trap led to an interface trap: The consultant assumed responsibilities that had not been agreed upon in the contract because the project would otherwise not have functioned. However, in doing so, he encountered difficulties in his accounting report because he now had to include services, which, according to both the contract and the rules of procedure, he should not have rendered. Maintaining the diplomatic protocol was ultimately deemed more important than the reason for initiating the entire game in the first place.

This shift in significance has the following background: The O script of development cooperation demands self-determination for all participants. A call for self-determination would be superfluous here, however, if it were not based on the assumption that the transfer of models was doomed to failure. A demand for self-determination only makes sense if one is convinced that successful transfer presupposes a more or less radical translation. This means, however, that the project-executing agency's frame of reference is different—otherwise there would be no need to translate the models. Conversely, anticipating, planning, and calculating would make no sense if there were not an epistemic link and an appropriate metacode connecting the frames of reference.

This dilemma of difference proves to be yet another rendering of the basic paradox of objectivity: No description of the world can claim validity outside of its own frame of reference insofar as any proof of validity always presupposes that particular frame of reference. Regardless of this, every description of the world assumes differences from other descriptions of the world, differences that cannot even be identified if no common, overarching frame of reference is presupposed. In dealing with this paradox in the realm of development cooperation, our interlocutors have observed a heightened sensitivity to differences between frames of reference. On the one hand, these differences constitute the obvious starting point of the entire matter (for what else could development cooperation do?) and they become painfully obvious in the course of practical translation work. On the other hand, in accordance with the diplomatic protocol and the technical game, they are translated into questions of loans, training, technology, and expertise, and in this way made invisible.

During the inception phase of the OIP (March 1996), Martonosi repeatedly insisted that the blueprint approach was disadvantageous in particular for project-executing agencies because this restricted their self-determination. He emphasized instead the advantages of the process approach, in which the model can be adapted to the local circumstances and local knowledge through participation and project ownership. A year and a half later, during the endless debates about the customer data that took place in the fall of 1997, Mbiti, the director of the waterworks in Baridi told Martonosi he had recognized from the very beginning that Martonosi's insistence that the process approach ostensibly worked in the interests of project-executing agencies in Ruritania was a ruse. Martonosi's ulterior motive, Mbiti continued, was that the professional incompetence of the African partners would lead them to agree to something as vague as the so-called process approach, which over the course of time would then give the consultants free rein to do whatever was in their own interest. Thus while Martonosi was convinced that he was intervening to create more maneuvering room in order to prevent yet another blind transfer of models that would ultimately ruin the project, he was accused of introducing a cultural difference into the game in order to reap advantages from it. In other words, Martonosi was attributed a position that was the diametrical opposite of that which brought him to the project as an organizational anthropologist in the first place—originally on the initiative of several NDB staff members. The remarkable thing about this disillusioning experience, however, is that cultural difference was rejected not by the center of calculation, but instead by the Ruritanian side.

In 1993, the World Bank published new guidelines for project strategies entitled *Getting Results: The World Bank's Agenda for Improving Development Effectiveness*. The central thrust of this agenda was that in order to increase the chances of success, it was essential that the borrower really wanted the financed measures; over the course of project implementation, project-executing agencies must adopt the projects as their own. To this end, they must be able to shape the projects in a definitive way, so it is essential that the necessary maneuvering room be created through the process approach. The World Bank publication summarizes these arguments under the catchword "project ownership." There is an implicit assumption in the study that is decisive for understanding this argument: The reason why project-executing agencies in the past did not follow calls to adopt projects as their own lay in the paternalism of the World Bank and its rigid guidelines. It is simply presumed without question here that this paternalism runs contrary to the interests of project-executing agencies, which must have a natural tendency toward self-determination.[6]

The key to understanding the issue lies in the fact that one of the Ruritanian project-executing agents regarded Martonosi's plea for more leeway in implementing the project to be a cunning tactical maneuver. Expanding the translation chains with centers of calculation and the ensuing spread of basic cultural technologies for monitoring and controlling from a distance is specific to modern societies with their distinct institutional apparatus. This characteristic marks the distinction between modern societies and those societies that qualify as "borrowers" in development cooperation. Development measures center around the attempt to introduce procedures of investigation and control from a distance as well as technologies of trust that are independent of local communitarian approval. If project-executing agencies now sacrifice precisely this distinction on the altar of the right to equality, they thereby negate the raison d'etre of development cooperation and build on sand—or rather on the wrong lists. In doing so, the recipients of development aid ultimately rob themselves of the chance to assume any perspective other than that of the hegemonic centers of calculation of the US–European world.

The ethnographic reports of this study have repeatedly demonstrated that project-executing agencies do not use the O script of development cooperation—which attributes central responsibility to them—as a means of demanding those aspects of development cooperation that have been factored out by the metacode of the technical game. The periphery's hasty approval of the guidelines drawn up by centers of calculation allows those centers to maintain their naive belief that their own technologies of investigation and control from a distance are merely neutral and objective instruments. For the centers of calculation, the metacode is not a pidgin trade language that follows a certain minimizing strategy capable of modification, but an objective descriptive language that is grounded in facts and consequently indisputable. If this is the case, however, the hegemony of development banks is then ultimately grounded in the fact that project-executing agencies participate in the technical game according to the unconditional aegis of the emancipation narrative for the simple reason that they believe this is the only way they can maintain their own autonomy. As a result, the most important corrective to development cooperation is lost: the recognition that a cultural difference exists.

Code Switching : Metacode—Cultural Code

The preceding sections have indicated that the metacode is indispensable for two reasons. First, centers of calculation can do their work only with the assistance of the metacode because their technical game is based on this code—and without centers of calculation, development cooperation would be impossible. Second, the inevitably humiliated recipients of development aid can maintain the reciprocity of perspectives only through the metacode—and without this reciprocity of perspectives, there would be no cooperation. Players, however, do not speak exclusively in the metacode in the trading zones of development cooperation (i.e., where questions marks appear in figure 6.5). Instead, the following distinction can be observed: Officially it is assumed that everyone accepts the metacode simply because it is valid. Unofficially, however, players assume both before and after negotiations that the other party (whoever that might be) is actually incapable of speaking in the metacode and just pretends to do so.

During negotiations in the metacode, all statements are regarded as true. This means that claims of objectivity cannot be questioned in principle without this leading to the termination of negotiations. Only against the backdrop of this premise can players argue about individual errors and deceptive maneuvers in order to arrive at an agreement of some kind. This premise, however, is frequently abandoned before and after negotiations, when, for example, the members of one party have gathered among themselves. They criticize the other party by casting doubt as to whether the latter's statements are relevant and factual. This criticism implies that the statements are grounded only in the other party's frame of reference and not in the reality of the world. In other words, these constructs are valid only relative to the culture of that particular party. This can be called a *cultural code*, in distinction to the ostensibly objective metacode.

In the course of our project, there were four major instances in which someone was accused of employing a cultural code. From the perspective of the three directors of the Ruritanian waterworks, customers were trapped in the culture of the socialist regime and therefore incapable of recognizing the necessity of paying for the treatment and supply of drinking water. From the perspective of the project consultant, employees at the waterworks were indeed able to use writing, forms, and files in a purely technical sense, but were unable to understand the sense and purpose of procedural objectivity. From the perspective of the project anthropologist Martonosi, any player who incorrectly ascribed an objective validity to the metacode (i.e., correspondence theory) and did not recognize it as a basic technique of hegemonic centers of calculation was speaking in the progress narrative, itself a cultural code. From Drotlevski's perspective, Martonosi had been deceived by the (constructivist) illusion that in negotiations and decision making it was possible to abstain from making assertions that were attributed objective validity and thus to establish a new (postmodern) order of business in which only various cultural codes existed. The "culture argument" is always brought in when players, with whom we either want to or have to cooperate, do not behave as we had hoped or as we regard as rational. In other words, culture comes in when we think they are wrong.

The initial observation regarding the proposal to abstain entirely from using the metacode because it is the actual source of failed cooperation under conditions of heterogeneity was in fact convincing. The naive belief (chapters 1 and 2) that it is possible to engage in development cooperation as a technical game, within the framework of which loans, technology, and expertise are transferred, renders invisible the fragile translation chain (figure 6.5), which is difficult to hold together behind the facade of the data highway (figure 6.4) and the machine metaphor (figure 6.3). This virtually eliminates any chance of correcting errors because it excludes the differences between the frames of reference employed by the cooperating parties, even though these remain the most significant sources of error.

However, Martonosi's constructivist proposal to treat the metacode as simply one of the many cultural codes (chapter 3) is not convincing as such. In the first place, his proposal ignores the conditions of its own possibility: Just as adherents of the metacode (otherwise called realists or objectivists) claim that the external world exists independent of human observers, critics of this position implicitly claim that the metacode's frame of reference, which is supposed to form the foundation of that code (e.g., the progress narrative), exists independent of their observations. Without such an assumption, it would be impossible even to formulate a constructivist argument in the first place: To assert the proposition "The metacode is only one of a number of cultural codes," the speaker must formulate this proposition itself in the metacode, the very possibility of which he or she is contesting. This irresolvable paradox demonstrates that metacode and cultural code do not represent mutually exclusive alternatives, but are instead the obverse sides of a single basic premise. In fact, the reports about the negotiation processes in our OIP indicate that players alternate back and forth between metacode and cultural code depending on the situation, without ever seeking to reconcile the resulting inconsistencies.

Second, the constructivist position also overlooks the indispensability of a metacode that claims absolute objectivity within the framework of ongoing negotiations. When cooperation is required under conditions of heterogeneity, players are forced—even against their own better knowledge—to reduce everything ultimately to a single dogged question: Is it correct or not? They cannot succeed in asserting a viewpoint while admitting at the same that this viewpoint is valid only relative to a particular frame of reference, which is itself merely relative and might just as well be altered. Finally, the constructivist position ignores both the significance of the metacode as a pidgin trade language for trading zones and its significance as the language of reciprocity.

Nevertheless, it is important to bear in mind a crucial distinction. The absolute conviction that the metacode is objectively correct (i.e., correspondence theory) is not identical with the relational conviction that the metacode is an inevitable presupposition for cooperation under conditions of heterogeneity. This relational conviction is the prerequisite for *reflection*, which players can engage in during pauses in negotiations, in order to correct errors not only by improving facts and figures, but also on a more fundamental level by reviewing their own frames of reference. In other words, to engage in reflection means to consider the objectivist metacode during pauses in negotiations as a cultural code and to review the basic cultural assumptions that surreptitiously flow into it. If it is impossible to replace the metacode with a cultural code, it is both possible and advisable to alternate continuously between the two in order to avoid being trapped by one's own blind spots. This recalls the elementary imperative of anthropology: to learn to see through the eyes of others, even if the other actors in this play with changing perspectives often do this unwillingly and rather poorly.

Notes

Prologue

1. Hobsbawm, *The Age of Extremes* (1995).

2. On the term "organizational field," see DiMaggio and Powell, "The Iron Cage Revisited" (1983). On the term "field" in general, see Bourdieu, "The Logic of Fields" (1992). The term "arena" (along with "social world" and "negotiation") is used in the present text in the same way it is used in: Strauss, *Negotiation* (1978a); Strauss, "A Social World Perspective" (1978b); Strauss, *Continual Permutations of Action* (1993); and Clarke, "Social Worlds / Arenas Theory as Organizational Theory" (1991).

3. Cited in Escobar, *Encountering Development* (1995), p. 3.

4. Quotation from Pinzer, et al., "Mein Maßstab: 'Das Lächeln eines Kindes.'" (1996). The shift in development rhetoric to an emphasis on ownership, participation, and good governance is one of the issues investigated in Mosse and Lewis (eds.), *The Aid Effect* (2005).

5. Quotation from Gourevitch, "Forsaken: Congo Seems Less a Nation than a Battle-field for Countless African Armies" (2000), p. 55. William Easterly, an economist at New York University and formerly at the World Bank, and one of the most ardent critics of large-scale development programs, has come to similar conclusions; see Easterly, *The White Man's Burden* (2006).

6. According to the Development Assistance Committee of the Organization for Economic Cooperation and Development (DAC/OECD), in 2006 the United States was the largest donor of "official development assistance" at $23.53 billion. DAC/OECD reports that the next largest donor was the United Kingdom ($12.46 billion). The UK was followed (in rank order) by Japan ($11.19 billion), France ($10.60 billion), Germany ($10.43 billion), the Netherlands ($5.45 billion), Sweden ($3.95 billion), Spain ($3.81 billion), Canada ($3.68 billion), Italy ($3.64 billion), Norway ($2.95 billion), Denmark ($2.24 billion), Australia ($2.12 billion), Belgium ($1.98 billion), Switzerland ($1.65 billion), Austria ($1.50 billion), Ireland ($1.02 billion), Finland

($0.83 billion), Greece ($0.42 billion), Portugal ($0.40 billion), Luxembourg ($0.29 billion), and New Zealand ($0.26 billion). Source: http://www.oecd.org/dataoecd/ 7/20/39768315.pdf (accessed on January 12, 2008).

7. A frequently cited example is Mali, which loses significantly more through US subsidies to American cotton farmers than it receives in development aid. The cotton subsidy case is so blatant that the World Trade Organization has demanded that the US government change this policy; see *New York Times*, September 1, 2004.

8. See Pandolfi, "Contract of Mutual (In)difference" (2003) and Pandolfi, "La zone grise des guerres humanitaires" (2006). The new trends that link development policy to security policy and to a new understanding of "global governance" were the topic of a conference held at the School of Oriental and African Studies (SOAS) in London in 2003; see the subsequent publication by Mosse and Lewis, *The Aid Effect* (2005). Mark Duffield, "Getting Savages to Fight Barbarians" (2005), offers an excellent overview of the shift from geopolitics to biopolitics, from "native administration" to the link between development and security. James Ferguson examines the links between globalization, neoliberalism, and governance in Africa after the end of the cold war; see Ferguson, *Global Shadows* (2006). Christophe Bonneuil examines the links between science and the state in Africa, and he shows how particularly between the 1930s and 1970s the colonial and postcolonial making of modern African states was a matter of experimentation with new forms of bringing together people, natural environments, and technologies in large development schemes; see Bonneuil, "Development as Experiment" (2001). In the present text, I attempt to show how the "experimentalization" of development changed under neoliberal conditions in the 1990s.

9. For a critical analysis of this trend see Jeremy Gould, "Timing, Scale, and Style" (2005).

10. On the Thomas theorem, see Merton, "The Self-Fulfilling Prophecy" (1949/1957). The exact beginnings of the post-Mertonian dispute can be identified only by drawing a relatively arbitrary line, for example, by providing the following list of names: David Bloor, Jacques Derrida, Nelson Goodman, Jean-François Lyotard, Willard van Orman Quine, and Richard Rorty in philosophy; in the social sciences and humanities, Jerome Bruner (psychology), Murray Edelman (political science), Michel Foucault (cultural history), Clifford Geertz (anthropology), Thomas Kuhn and Steven Shapin (history of science), Barry Barnes, John Law, and several others (sociology of science), D. N. McCloskey (economics) and Hayden White (history). For measured analyses of the misunderstandings and irritations between the two camps, see Smith, *Belief and Resistance* (1997) and Zammito, *A Nice Derangement of Epistemes* (2004).

11. See Clifford and Markus, *Writing Culture* (1986). In the present book I do not intend to comment explicitly on either the writing of culture or the writing culture debate. I believe enough has been written on that issue. A surplus of reflections about reflexivity quickly becomes sterile; see Pinch, "Reservations about Reflexivity and

New Literary Forms" (1988). We need new ethnographies that are as vigorous and vibrant as classic ethnographic studies. This is the task that I set for myself in writing this book. In 1986, the year that *Writing Culture* was published, another volume appeared that was equally influential, *Power, Action and Belief*, edited by John Law. The latter was extremely important for the present study. I also profited from anthropological insights into the politics and poetics of ethnographic writing. Although these insights are not explicitly elaborated in the body of this text, they did inform my textual strategy.

12. The anthropological turn to the citadels of modernity (organization, science, and technology) was initiated outside of the discipline itself, in particular by the history, philosophy, and sociology of science (from Ludwig Fleck, Karl Mannheim, and Max Scheler through Robert K. Merton, Thomas Kuhn, Barry Barnes, and David Bloor up to the current debates for instance in the journal *Social Studies of Science*). Some of this literature has been used in this book; here I refer only to a magnificent outline of this problem by Isabelle Stengers, *The Invention of Modern Science* (1993/2000).

13. Quotation from Johnson, "Conflicting Images and Realities in Project-Planning" (1985), p. 351. Although the year 1985 is located in the not-all-too-distant past, the mention of telexes in the citation make it sound like a quotation from antiquity, as younger readers may never have seen or even heard of a telex. This, however, demonstrates at the same time that the global network existed even before the Internet. James Ferguson makes the same point about the necessity of transcending anthropological localism in order to grasp the translocal processes that have an enormous impact on the social order in African localities; see Ferguson, *Global Shadows* (2006), p. 3 specifically, and generally throughout the book.

14. On the notion of distributed agency, see Garud and Karnøe, "Bricolage versus Breakthrough" (2003). On the inclusion of nonhuman actors, see Callon, "Some Elements of a Sociology of Translation (1986).

15. Here I mean, of course, all of Luhmann's work, but the point is presented well in Luhmann, "Sthenographie und Euryalistik" (1991), here p. 60. Unfortunately and curiously, Niklas Luhmann is less well known to the Anglo-Saxon scholarly world than his European colleagues of similar import from the second half the twentieth century, such as Pierre Bourdieu and Michel Foucault. In January 2008, the catalog of the Cornell University library system listed 427 titles relating to Foucault, 140 to Bourdieu, and 104 to Luhmann (15 of which are English translations). Although my own approach does not adopt Luhmannian systems theory, I do make use of a number of his insights throughout the book.

My discussion of the epistemological question of a so-called deflationary strategy in anthropological investigations has been drawn from the field of science and technology studies (STS). A good overview of this now broad field is presented in the first chapter of Pickering, *The Mangle of Practice* (1995), pp. 1–34. My position here can essentially be traced to Bruno Latour, who makes *translation* or *mediation* his starting point in overcoming the helpless opposition of reality and representation (I address this issue in more detail in the following section of the prologue). Similar to systems theory and deconstruction, Latour intends to offer an alternative to the aporias of the classical discourse of truth. See in general Latour, *We Have Never Been Modern* (1993). I do not, however, share Latour's optimism about the possibility of avoiding objectivism in "real life." My investigation concurs more with Luhmann, who presumes that we have to suspend doubt when confronted with the necessity of making decisions. Wittgenstein formulated this aptly in *On Certainty* (1969): "And doubting means thinking (Proposition 480). . . . The reasonable man does not doubt certain things (Proposition 220). . . . A doubt without an end is not even a doubt (Proposition 625)."

16. Goffman, *Frame Analysis* (1974/1993).

17. The anthropology of laboratory work in the natural sciences appears to take place under less precarious circumstances—presumably because both sides regard it as an issue of science and not of politics, power, and money, as is the case in the present study. See Knorr Cetina, *The Manufacture of Knowledge* (1981); and Law, *Organizing Modernity* (1994). In contrast, Hugh Gusterson, who investigated the laboratories of the atomic weapons industry in the United States, reports comparable problems in observing nuclear scientists; see Gusterson, *Nuclear Rites* (1996). Gusterson resolved the problem by remaining at a distance and working primarily with interviews.

18. In Vladimir Nabokov's novel *Pnin*, this quadrangular configuration of authorities assumes the following form: There is the empirical author Nabokov (first authority), and N., the first-person narrator, author, and observer (second authority). The figure of Pnin plays the role of the skeptic (third authority), who has difficulty adjusting to life in the United States. The role of the believers (fourth instance) is played by the diverse representatives of US society, whom Pnin encounters throughout the novel. On this issue, see Adorno, "The Position of the Narrator in the Contemporary Novel" (1958/1991).

19. This particular form of indirect representation is similar to that employed by Latour and Woolgar in *Laboratory Life* (1979/1986), the first ethnography of a scientific laboratory.

20. The distinction between tactics and strategies that I follow here has been drawn from Michel de Certeau, *The Practice of Everyday Life* (1980/1984).

21. This particular approach and the interest in "interstitiality" has also been inspired by Joerges and Shinn, "Research-Technology in Historical Perspective" (2001). The issue of mediation and the various social figures who function as brokers is crucial within the field of the anthropology of development. Bierschenk and a number of his colleagues have drawn attention to the often invisible and therefore easily underestimated work of brokers at the local village level in Africa; see Bierschenk, Chauveau, and Sardan, *Courtiers en développement* (2000). Analogous to their enterprise I have shifted the attention to another field of largely invisible "courtage" in development, where the key figure is the international consultant.

22. On the distinction between the terms "plot" and "story," see Eco, *Six Walks in the Fictional Woods* (1994). The story is the course of the events as they occur (the *Odyssey*, for example, begins with the departure from Troy). The plot is the construction of the narrative that has been selected in order to tell this story (Homer's text begins with Odysseus as a captive of the nymph Calypso and then reconstructs the course of events through flashbacks).

1   The Solid Ground of Development Policy

1. According to OECD statistics, aid funds to sub-Saharan Africa increased from 9 billion dollars in 1985 to 18 billion dollars in 1995, stagnated around 13 billion until 2001, and then began to rise sharply between 2002 and 2006 from 18 to 40 billion dollars. The 2006 increase to 40 billion is related primarily to attempts to control the HIV crisis (see http://stats.oecd.org/wbos/default.aspx?DatasetCode=ODA _RECIPIENT_REGION, data extracted on December 25, 2007).

2. See William R. Scott, *Organizations* (1981/1987).

3. See Arrow, "The Economics of Agency" (1985/1991).

3   The Disclosures of Science

1. Luhmann, *Funktionen und Folgen formaler Organisation* (1964/1995), p. 110.

2. On the distinction between first-order and second-order observation, see the concise formulation in Luhmann, "Sthenographie und Euryalistik" (1991), especially p. 63.
    Within the official frame of reference in development cooperation, anthropology enters as the voice of developing societies. For an initial overview of what is called "local knowledge," see Sillitoe, "What, Know Natives?" (1998).
    The realization that it is necessary to shift the focus to the US–European side, to the organizations of development cooperation, and to the discourse of development, however, is not entirely new; see, for example, Hoben, "Anthropologists and Development" (1982) and Bennett and Bowen (eds.), *Production and Autonomy* (1988). One author who followed this suggestion early on is Quarles van Ufford; see Quarles van Ufford, "The Myth of Rational Development Policy" (1988a), "The Hidden Crisis in Development" (1988b), and "Knowledge and Ignorance" (1993). For the perspective of a practice-oriented political scientist, see Elshorst, "Organisation und Entwicklung" (1990). For a systems-theory perspective, see Hanke, "Weiß die Weltbank, was sie tut?" (1996). Richard Harper presented an ethnography of documentation within the IMF. Similar to the present study, he concentrates on what he calls the "anthropology of documents"; in contrast to this book, however, he limits himself to the processes within a single organization and, unfortunately, does not situate his interesting observations within a wider field of practices. See Harper, *Inside the IMF* (1998). The most influential authors in this regard are Arturo Escobar, "Power and Visibility" (1988) and James Ferguson, *The Anti-Politics Machine*

(1990). A few years after the publication of the German original of this book, *Weit hergeholte Fakten* (Rottenburg 2002), Tania Murray Li published her excellent monograph *The Will to Improve* (2007) on the emergence of World Bank neoliberal strategies in Indonesia.

3. As is well known, one of the central focuses of Max Weber's work is the rationalization of the world; see Weber, "Religious Rejections of the World and Their Directions" (1920/1948), pp. 323–362, in particular pp. 331–333. In his reflections on the process of rationalization, Weber resorts to the distinction between internal and external morality; see primarily Weber, *General Economic History* (1923/1927), "The Evolution of the Capitalist Spirit," pp. 352–369, here p. 356. A further examination of this issue is Nelson, *The Idea of Usury* (1969). Georg Elwert has developed a phenomenology of (what I call here) the external world in the complex societies of developed countries, which is oriented around the concepts of expansive venality, the commodification of social relations, the empowerment of the economy, and the remoralization of politics; see Elwert, "Ausdehnung der Käuflichkeit und Einbettung der Wirtschaft" (1987a). On this issue in general, see Eisenstadt, "Anthropological Studies of Complex Societies" (1961); Polanyi, "The Economy as Instituted Process" (1957/1971); Münch, *Struktur der Moderne* (1984); and Powell, "Neither Market nor Hierarchy" (1990).

4. On the distinction between personal trust and system trust, see Luhmann, *Trust and Power* (1979). On the consequences of eroded trust, see Waldmann, "Anomie in Argentinien" (1996).

5. See Münch, "Max Webers 'Anatomie des okzidentalen Rationalismus'" (1978), especially pp. 241–242; Hirschman, "Social Conflicts as Pillars of Democratic Market Society" (1994); and Dubiel, "Der Fundamentalismus der Moderne" (1992).

6. On this issue, see Gellner, *Nations and Nationalism* (1983); Polanyi, *The Great Transformation* (1944/2001), in particular chapter 14, "Market and Man," pp. 171–186; and Thompson, *The Making of the English Working Class* (1963/1980). On the logic of the self-dissolution of the Western welfare state, see Buchanan, *Liberty, Market, and State* (1985).

7. A systematic project evaluation for the United States Agency for International Development found that "inappropriate time phasing of project activities" is one of the nine most important causes of project failure; see Gow and Morss, "The Notorious Nine" (1988), p. 1409. What this article examines in the realm of organization and development cooperation has already been described repeatedly in systems theory and neoinstitutional organizational research. See Luhmann, *Funktionen und Folgen formaler Organisation* (1964/1995), p. 110. Luhmann also refers to several precursors of this observation in the domain of general sociology, in particular to Erving Goffman with his focus on the "dilemma of expression versus action," as well as Charles Perrow and Amitai Etzioni. An influential argument regarding this point has also been presented by Meyer and Rowan, "Institutionalized Organizations" (1977/1991).

8. On the paradox of organizations, see March and Simon, *Organizations* (1958); Blau, *Bureaucracy in Modern Societies* (1956); Meyer and Rowan, "Institutionalized Organizations" (1977/1991), pp. 56–60; Brunsson and Olsen, *The Reforming Organization* (1993); Weick, *Sensemaking in Organizations* (1995); Czarniawska-Joerges, *The Three-Dimensional Organization* (1993); and Czarniawska, *Narrating the Organization* (1997). Directly related to the present text, see Hanke, "Weiß die Weltbank, was sie tut?" (1996).

9. Meyer and Rowan, "Institutionalized Organizations" (1977/1991), p. 55.

10. Within organizational research, the Weberian concept of rationalization has been revised on the basis of this distinction between representation and practice. On the "loose coupling" that results from this, see especially Weick, *The Social Psychology of Organizing* (1969/1979), pp. 335–336. For an example that should interest anthropologists, see Weick and Roberts, "Collective Mind in Organizations: Heedful Interrelating on Flight Decks" (1993). It has been argued from a neoinstitutional perspective that certain organizational models become established not because they are rational and therefore efficient, but because they are regarded as rational; see DiMaggio and Powell, "The Iron Cage Revisited" (1983).

On the connection between loose coupling in processes of formal-rational organizing and the philosophical question of truth or reality (drawing explicitly on Wittgenstein, Brillouin, and Serres), see Lyotard, *The Postmodern Condition* (1979/1984), pp. 55–56. Within communications theory as formulated by Luhmann, the phenomenon of loose coupling is considered constitutive for all human understanding (*Verstehen*): For understanding to occur at all, the recipient of information must be able to distinguish between information and message. If information and message were identical—i.e., tightly coupled—then it would be impossible for the recipient of information to identify the selection made by the person communicating the information. This would render superfluous the role of the communicator and as a result there would be nothing more to understand (as in machine-to-machine communication); see the chapter "Communication and Action," in Luhmann's *Social Systems* (1984/1995), especially pp. 139–157. In identity theory, Goffman's observations on role distance and George Herbert Mead's distinction between "I" and "me" correspond to the phenomenon of loose coupling. See on the latter Krappmann, *Soziologische Dimensionen der Identität* (1969/1975), especially pp. 132–173. See also Sartre on the necessity of bad faith for identity formation in Sartre, "Bad Faith" (1943/1956).

11. Along with deregulation and privatization, the loss of trust in scientific knowledge has contributed to the erosion of the aura of professors as being beyond any kind of oversight, which in turn has led to the end of the era of the professor; on this see Lyotard, *The Postmodern Condition* (1979/1984), p. 53.

12. Meyer and Rowan, "Institutionalized Organizations" (1977/1991), p. 57. The following reflections on the characteristics of institutional organizations are oriented around Meyer and Rowan.

13. Although this argument is evident in the neoinstitutionalism of organizational research (Meyer and Rowan, "Institutionalized Organizations" [1977/1991], pp. 56–60), Michael Power has provided an authoritative exegesis of it in his study, *The Audit Society* (1997). The arguments I present here are based on Power's book. On the connection between deregulation, science accounting, and peer review, see Fuller, "Toward a Philosophy of Science Accounting" (1994), here pp. 257–258.

14. A clear example of this critique is presented in Gow and Morss, "The Notorious Nine" (1988), p. 1412, where the "blueprint approach" is traced back to the interests and dominance of Western experts—an oversimplification that I attempt to refute with this study.

15. On this fundamental critique of conventional social science since Durkheim, see Latour, "The Power of Association" (1986). Niklas Luhmann has made the same point from the perspective of systems theory repeatedly since the 1970s and presented an excellent summary in his emeritus lecture; see Luhmann, "Was ist der Fall?" (1993).

16. Gellner, *Reason and Culture* (1992), pp. 3, 10.

17. The considerations presented here on the connections between progress, rationalization (rationalism/objectivism), and science can be traced primarily back to Max Weber, as noted earlier; see Weber, *The Protestant Ethic* (1904/2002); see also Gellner, *Nations and Nationalism* (1983), especially pp. 19–24.

18. Citations are from Gellner, *Nations and Nationalism* (1983), pp. 19–24.

19. See Gellner, *Nations and Nationalism* (1983), pp. 32–33. On mobility and its consequences in the sense addressed here, see also Sennett, *The Corrosion of Character* (1998).

20. Cited in Lyotard, *The Postmodern Condition* (1979/1984), p. 29. The preceding argument about the outsourcing of knowledge to special institutions as well as the subsequent reflections about objectivity are also based on Lyotard's *The Postmodern Condition*, pp. 7, 8–9, 25, 37–41.

21. Discourse in the social sciences and the humanities over the past twenty years about how reality is always a constructed reality often gives the impression that the protagonists believe they have discovered something entirely new. And the counterarguments are occasionally even more simplistic—tapping one's fingers on a table to demonstrate its reality, or telling one's adversary that if everything is negotiable he should jump out the window and negotiate the laws of gravity. It is perhaps helpful here if we bear in mind, again, that this debate is as old as humanity itself. If we place the issue in a historical context, it is easier to identify the contemporary points of contention. On the ubiquity of this dispute about the nature of reality, see, for example, Hegel, *The Phenomenology of Mind* (1807/2003), pp. 44–53, especially p. 46; Blumenberg, *Wirklichkeiten, in denen wir leben* (1981/1996). For the social sciences, Weber's "'Objectivity' in Social Science and Social Policy" (1904/1949)

remains largely unchallenged, although the continuing debate seems to keep forgetting this.

Contemporary constructivism/deconstruction attempts to move beyond the older, phenomenological-hermeneutic understanding of the unattainability of reality, although it is not easy to detect the precise point or points of difference. Important works for the present study include: Lyotard, *The Postmodern Condition* (1979/1984), pp. 24–27, 37–41; and Rorty, *Philosophy and the Mirror of Nature* (1979). Rorty draws upon the work of Wittgenstein, Heidegger, and Dewey, continuing their critique of knowledge as an accurate representation of external reality. For a trenchant summary of his argument, see Rorty, "Representation, Social Practice, and Truth" (1991b). It appears plausible that these authors (in particular Lyotard) identify a connection between their own critical ambitions and an actual social condition—the transformation of labor—without, however, reducing one to the other. This would mean that current constructivism seeks answers to pressing contemporary social issues, as has already been argued about the works of Marx, Durkheim, and Weber. Luhmann, too, is not concerned solely with progress within theory, but also with the development of social differentiation; see Luhmann, *Die Wissenschaft der Gesellschaft* (1990/1994), pp. 616–701. In the narrower sense, my argument here is oriented around Latour's "The Politics of Explanation" (1988), "Clothing the Naked Truth" (1989), and Latour, "The Force and Reason of Experiment" (1990), pp. 48–79, here especially pp. 63–71.

One contentious difference between phenomenological-hermeneutic constructivism (Weber) and contemporary (de)construction (Lyotard, Rorty, Luhmann) lies in their respective views on how the connection between knowledge and power should be addressed, if this connection is in fact inevitable. The former (for example, Gellner) resolves this paradoxical issue by ignoring it and presenting "objective analyses." The latter (Lyotard, Rorty) resolve it by rejecting any and all claims to objectivity that do not immediately expose themselves as particular. In doing so, they displace the basis for justice and solidarity. More recent research on science (in particular Latour's essays just noted) has assumed an intermediate position, which attempts to circumvent the problem of juxtaposing "word" and "world" by conceiving of *representations as translation* practices that are then placed in an ethnographic context. On the connection between power and representation (in the sense of political representation and scientific depiction), see the groundbreaking article by Callon, "Some Elements of a Sociology of Translation" (1986).

22. In lieu of any ultimate foundations, Habermas focuses more on the possibilities of the emancipation narrative pulling itself up by its own bootstraps. In this sense, he views consensus as the agreement that human beings reach in free and uncoerced discourse on the basis of plausible experience; on this see Habermas, *Knowledge and the Human Interests* (1968/1971); and Habermas, *Justification and Application* (1993). For Luhmann, who offers autopoiesis as a bootstrapping method, consensus becomes primarily an object of administrative procedure; on this see Luhmann, *Legitimation durch Verfahren* (1969/1973). Lyotard, in contrast, attempts to develop an idea and a practice of justice that are not tied to consensus (which he believes has become impossible). According to Lyotard, this requires acknowledging

the heteromorphy of language games and the resulting repudiation of terror; on this see Lyotard, *The Postmodern Condition* (1979/1984), pp. 60–67. Rorty's and Latour's arguments on this are similar to those of Lyotard; see Rorty, *Objectivity, Relativism, and Truth* (1991a); and Latour, *We Have Never Been Modern* (1991/1993). For the counterposition, see Gellner, *Reason and Culture* (1992), especially p. 167.

23. See the much discussed and controversial contribution by Robin Horton, which argues that even assuming a universal rationality does not mean we have to agree on a single uniform conception of the world (Horton, *Patterns of Thought in Africa and the West* [1993]); the book is an anthology that also contains two older essays from 1967 and 1988.

24. I have already addressed this expanded conception of translation several times, in particular in the prologue and the extensive notes 21 and 22 above. The term translation *chain* has been taken from Latour, *Science in Action* (1987), pp. 132–144. To a certain extent, focusing on translation chains shifts the point of departure of the analysis back a step: Identities, actors, interests, objectives, organizations, and larger formations (for example, structures and fields/arenas) are not introduced as fixed points in order to explain the course of actions. Instead, the standpoint is assumed that if actions establish connections, that is, if they produce translations, then chains are generated that are in turn woven into networks. Only in the course of these practices do so-called fixed points emerge. These are certainly not insignificant for the course of further practices, but neither do they determine them. On this, see Latour, "The Power of Association" (1986); and Latour, "On Actor-Network Theory" (1996). For the applicability of the concept of a translation chain in anthropology, see also Czarniawska and Joerges, "Travels of Ideas" (1996). For an example of a case study, see Czarniawska, *A City Reframed* (2000).

25. Murray Edelman has examined political discourse and shown how its inner logic omits any questions that could undermine the foundations of the existing political order as a whole; see Edelman, *Constructing the Political Spectacle* (1988), pp. 27–29. For an examination of this phenomenon at the level of organizations, see chapter 8 ("Darstellung des Systems für Nichtmitglieder") in Luhmann, *Funktion und Folgen formaler Organisation* (1964/1995), pp. 108–122, especially p. 114.

26. World Bank, *Getting Results* (1993), pp. 7, 12 (emphasis added). For an influential and practice-oriented article (published in the late 1980s) that argued that the participation approach would solve many problems of development cooperation, see Gow and Morss, "The Notorious Nine" (1988). For an overview of support provided by anthropology for this position, see Sillitoe, "What, Know Natives?" (1998).

27. See Lyotard, *The Postmodern Condition* (1979/1984), p. 64, footnote 222, which cites Orwell and recalls Watzlawick's classical reflections on "Be free!" and "Want what you want!"

28. On "defensive communication," see Scott, *Domination and the Art of Resistance* (1990).

29. See Lyotard, *The Postmodern Condition* (1979/1984), pp. 31–46, especially p. 44. Lyotard refers here to Luhmann, who examined the significance of the performativity of procedures in postindustrial society, a tendency that has become so extensive that to some extent the performativity of procedures even replaces the normativity of the procedures; see Luhmann, *Legitimation durch Verfahren* (1969/1973). Luhmann's reflections also refer to the well-known first chapter of Marcuse's *One-Dimensional Man* (1964). See also Dubiel, "Der Fundamentalismus der Moderne" (1992); Guattari, "Praktiken der Zukunft" (1994); and Joerges, *Technik—Körper der Gesellschaft* (1996).

30. Latour, *Science in Action* (1987).

31. Arturo Escobar has succinctly described the drawbacks of such an enterprise: "The local situation is inevitably transcended and objectified as it is translated into documentary and conceptual forms that can be recognized by the institutions. In this way, the locally historical is greatly determined by nonlocal practices of institutions." Escobar, "Anthropology and the Development Encounter" (1991), p. 667. Annelise Riles has presented a critical analysis of empowerment and capacity-building in her study on NGO networks in the Pacific region with regard to the Beijing "Women's Decade" Conference in 1995. She examines how the preoccupation of the women's organizations with documentation and forms, with what she calls aesthetics and style, became an end in itself: The aesthetics of the network became its substance. If an ulterior motive for this kind of aesthetics existed, it was to facilitate access to funding from donor agencies that expected certain forms; see Riles, *The Network Inside Out* (2001). In other words, the local is in some sense determined by the translocal, as Escobar says, but not necessarily in the sense that was intended by some dominant players, as I have attempted to demonstrate in this study (contra Escobar and Ferguson).

32. Scott, *Seeing Like a State* (1998) is a large-scale study of this issue. Scott, however, concentrates on the *consequences* of what I have called the technical game, without investigating the game itself. In doing so, he remains true to the division of labor that he subscribed to decades earlier with his colleague Charles Perrow at Yale, with Perrow focusing primarily on questions of how formal organizations work and Scott taking up the consequences of expanding the technical game. In this sense, I am attempting to bring together Perrow and Scott.

4 Interstitial Spaces

1. The distinction between a "narrative mode of knowing" and a "logo-scientific mode" comes from Jerome Bruner, who took up and further developed Lyotard's concept of narrative knowledge. In simple terms, this means that all efforts to make sense of a situation consist of ordering experience into a familiar narrative in order to orient oneself. The repertoire of narratives available to a person, a group, or a culture is hierarchical, so that certain master narratives (or metanarratives or key narratives) serve to prestructure smaller, more situational narratives. See also

Bruner, *Actual Minds, Possible Worlds* (1986); and Lyotard, *The Postmodern Condition* (1979/1984). Neoinstitutionalism, with its emphasis on the (Weberian) notion of legitimation and validity, has in principle been responsible for the transfer of the narrative/rhetorical approach to organizational research. Recent contributions on the subject include: Weick, *Sensemaking in Organizations* (1995); and Czarniawska, *A Narrative Approach to Organization Studies* (1998).

2. Cited from Gow and Morss, "The Notorious Nine" (1988), p. 1407 (emphasis added); the following quotation was taken from p. 1413 of the same source. On the statistics in the preceding paragraphs see: "Die Qualität der Weltbankprojekte verbessert sich," *Frankfurter Allgemeine Zeitung* (December 19, 1997).

3. The three quotes are taken from: Martin, "The Deconstruction of Development" (1998); Escobar, "Anthropology and the Development Encounter" (1991), p. 676; and Sillitoe, "What, know natives?" (1998), p. 203. As an example of the many moderate and conciliatory voices on the argument, see Hobart, "Introduction: The Growth of Ignorance?" (1993). A survey of the more hopeful voices is offered in the aforementioned Sillitoe, "What, know natives?". For a lively but less conciliatory example of a rural development project see Beck, "Entwicklungshilfe als Beute" (1990). The following two works reject in principle the possibility that development projects can change things for the better: Sachs (ed.), *The Development Dictionary* (1992/1997); and Scott, *Seeing Like a State* (1998). More recent publications have added weight to a skeptical perspective; see Mosse and Lewis, *The Aid Effect* (2005), and Easterly, *The White Man's Burden* (2006).

4. Gow and Morss, "The Notorious Nine" (1988), p. 1415. Another applied study that addresses the question of participation in a sophisticated manner is presented by Sülzer and Zimmermann (eds.), *Organisieren und Organisationen verstehen* (1996). However, precisely this study gives the impression that self-deception regarding the aporias of participation does not necessarily decrease as social scientific standards rise, as long as the dogma remains that development cooperation is possible in principle in the present form and that only minor modifications are necessary.

5. On the theoretical background of this argument, see the discussion in chapter 3 and, in detail, Latour, "Drawing Things Together" (1988/1990). From the symbolic interactionism perspective, see also Star and Griesemer, "Institutional Ecology, 'Translations' and Boundary Objects" (1989); and Fujimura, "Crafting Science: Standardized Packages, Boundary Objects, and 'Translation'" (1992). See also van der Sluijs et al., "Anchoring Devices in Science for Policy" (1998). On the concept of the "trading zone," see Galison, *Image and Logic* (1997), especially chapter 9. The standardized packages are generally not effective if the differences they are intended to link are too basic or contain implicit orientation patterns, such as: (1) definitions of reality, semantics, classifications (What is the case? What is it about? Who are we?); (2) definitions of priorities, values (What do we want? What does it come down to? What is significant? In view of what is something important?); and (3) definitions of plausibility, causalities (How can I know what applies? How are the things related?). If things become dissonant in transcultural interactions, it is because the implicit,

preconscious part of these orientation patterns (1–3) lets the explicit interpretations and actions diverge. On this see Elwert, "Kulturbegriffe und Entwicklungspolitik" (1996), p. 54. On the general pattern of evading differences see also Schiffauer, "Die Angst vor der Differenz" (1997).

6. The NUWA was transformed into UWASA (Urban Water and Sewerage Authority) in 1997, largely according to the ideas developed by S&P for the other three cities. In 2003, five years after the newspaper announcement, UWASA was actually privatized under the guidance of the World Bank and the IMF. This step was initially celebrated as a harbinger of water privatization for Africa but soon became very controversial, already ending in scandal in 2005, when the government abruptly abrogated the contract with the globally operating private companies.

7. Albert O. Hirschman described model transfer in development cooperation as pseudo-imitation as early as 1970: Development projects increase their legitimacy by being transferred as images and repetitions of successful endeavors in other places in the world, usually the industrialized, Western world; see Hirschman, *Development Projects Observed* (1967). For a general view of the relationship between effectiveness and the spread of organizational models, see DiMaggio and Powell, "The Iron Cage Revisited" (1983). With respect to our case, this means that the BOOT principle is being considered in Baharini because it was successful in Britain.

Harald Scherf speaks of economic fundamentalism that unfolds, suspiciously enough, precisely in the face of the collapse of socialist planned economy. He says that this fundamentalism is based mainly on forgetting a simple fact, namely, that all successful capitalist countries in the West have been concerned for more than a century primarily with restricting and regulating the scope of market mechanisms, rather than letting them run wild. On this see Scherf, "Fundamentalismus in der Ökonomie" (1992), in particular pp. 816–819. Thomas Petersen's critique of James Buchanan's work is also noteworthy; see Petersen, *Individuelle Freiheit und allgemeiner Wille* (1996). Michel Callon reversed the argumentation and showed how markets must first be made before they can develop their laws; see Callon (ed.), *The Laws of the Markets* (1998). Continuing along that line, see also Rottenburg, Kalthoff, and Wagener, "In Search of a New Bed" (2000).

8. On the World Bank, see the ethnographic essay by Enzensberger, "Billions of All Countries, Unite!" (1988/1997).

9. On a paradigm change that, since the 1970s, has shifted the debate about economic development from optimizing centralized planning to understanding an evolutionary process, see Weiss, "Changing Paradigms of Development in an Evolutionary Perspective" (1992). For a perspective on the social distribution of knowledge along lines that strongly support current development policies in the direction of deregulation and decentralization see Hayek, "The Use of Knowledge in Society" (1948/1996). Without delving into the point that too much planning obstructs the desired development, Gow and Morss argue empirically that the coordination problem discussed here is one of the nine notorious reasons for the failure of development cooperation; see Gow and Morss, "The Notorious Nine" (1988), pp. 1400–1402.

10. See Gow and Morss, "The Notorious Nine" (1988), p. 1403.

11. The generally unsuccessful attempts to deal with empowerment (in the context of decentralization and participation) are one of the "notorious nine critical problems in project implementation"; see Gow and Morss, "The Notorious Nine" (1988), pp. 1407–1408.

5   Trading Zones

1. Within the field of development cooperation and the anthropology relating to the field, mapping is not treated as a cultural practice that varies according to cultural areas and historical times. In contrast, cognitive anthropology and social studies of science and technology readily assume cultural differences in mapping and consider them to be of great interest; see for instance David Turnbull, *Masons, Tricksters, and Cartographers* (2000), especially chapter 3.

2. Rottenburg, "Social Constructivism and the Enigma of Strangeness" (2006).

3. The argument about the strength of an explanation is based on Latour, "The Politics of Explanation" (1988). For basic positions on writing, lists, tables, and recipes, see Goody, *The Domestication of the Savage Mind* (1977); Goody, *The Logic of Writing and the Organization of Society* (1986), chapters 2–4, pp. 45–170; Assmann, *Das kulturelle Gedächtnis* (1992); Assmann, "Lesende und nicht lesende Gesellschaften" (1994); available in English translation, Assmann, *Religion and Cultural Memory* (2006); Elwert, "Die gesellschaftliche Einbettung von Schriftgebrauch" (1987b). The theoretical text that Elwert substantiates from an anthropological perspective is Luhmann, *Social Systems* (1984/1995), pp. 157–163. On bureaucratic logic see also Don Handelman, "Cultural Taxonomy and Bureaucracy in Ancient China" (1995).

4. Borges, "The Analytical Language of John Wilkins" (1964), pp. 101–105, here pp. 103–104.

5. See the previous discussion of the distinction between written/scientific and oral/narrative knowledge in chapter 4.

6. For a general treatment of this see Pinch, "Toward an Analysis of Scientific Observation" (1985); and Porter, *Trust in Numbers* (1995). On the matter discussed here specifically, see Power, *The Audit Society* (1997).

7. The dialectic between the repression of indigenous knowledge through the introduction of various technologies of governance (such as procedural objectivity and forms of bureaucratic standardization to make things legible) and the enunciation of indigenous knowledge as a separate, endangered, and highly valuable form of knowledge is examined by Christophe Bonneuil, "Development as Experiment" (2001).

8. The UWEs' refusal to become involved in the project in the sense of the official conception of participation brings to mind Sartre's concept of "active passivity." The

running debate on Herman Melville's story *Bartleby, the Scrivener* accords with this concept. To the extent that resistance generally leads to compliance with the hegemonic rules of the game, this debate focuses on how it is possible to demonstrate resistance without supporting the hegemonic system. This is where the issue of active passivity comes into play; see Sartre, *Critique of Dialectical Reason* (1960/1991), and Beverungen and Dunne, "I'd Prefer Not To" (2007).

9. In 2006, a local newspaper reported that the waterworks of Baridi, Mlimani, and Jamala had become the national models for operating urban waterworks in Ruritania. The article stated that revenue collection for Jamala had tripled between 1993–94 and 2004–05, water was running 24 hours a day, and collection efficiency increased to 94% in 2004. At the same time, the number of employees per 1,000 points of sale had been reduced from 14 in 1997 to 8 in 2005.

6 Metacode—Cultural Code

1. Figure 6.1 has been adapted from Douglas and Wildavsky, *Risk and Culture* (1982), p. 5.

2. Figure 6.5 has been taken from Bruno Latour, who uses it to make a fundamental argument against the opposition of "world" and "language" as *adaequatio rei et intellectus* and asserts, in contrast, that reference as transversal reference is itself a quality of the translation chain; on this, see Latour, "The 'Pédofil' of Boa Vista: A Photo-Philosophical Montage" (1995). This issue has already been addressed in endnotes 15 (prologue), 21, 22, 24, and 29 (chapter 3).

3. Two colleagues have engaged in more in-depth investigations within the scope of the social studies of finance and the growing interest in the sociology of calculation; their research is directly related to my own work in number of ways. See Kalthoff, "Ökonomisches Rechnen" (2007) and Kühl, "Zahlenspiele in der Entwicklungshilfe" (2007); see also the joint publication Rottenburg, Kalthoff, and Wagener, "In Search of a New Bed" (2000).

4. See Schütz and Luckmann, *The Structures of the Life-World* (1973/1989), p. 5; and Schütz, *The Phenomenology of the Social World* (1932/1967), pp. 97–137. In another theoretical language this is called the "principle of charity"; see Davidson, *Inquiries into Truth and Interpretation* (1984/2001).

5. See Galison, *Image and Logic* (1997), especially chapter 9, "The Trading Zone: Coordinating Actions and Belief," pp. 781–844.

6. World Bank, *Getting Results* (1993).

References

Adorno, Theodor W. 1958/1991. The Position of the Narrator in the Contemporary Novel. In Theodor W. Adorno, *Notes to Literature*, vol. 1, 30–36. Edited by Rolf Tiedemann, translated by Shierry Weber Nicholsen. New York: Columbia University Press.

Arrow, Kenneth J. 1985/1991. The Economics of Agency. In *Principals and Agents: The Structure of Business*, edited by John W. Pratt and Richard J. Zeckhauser, 37–51. Boston: Harvard Business School Press.

Assmann, Jan. 1992. *Das kulturelle Gedächtnis: Schrift, Erinnerung und politische Identität in frühen Hochkulturen*. Munich: Beck.

Assmann, Jan. 1994. Lesende und nicht lesende Gesellschaften. Zur Entwicklung der Notation von Gedächtnisinhalten. *Forschung und Lehre* 1 (2): 28–31.

Assmann, Jan. 2006. *Religion and Cultural Memory: Ten Studies*, translated by Rodney Livingstone. Stanford, Calif.: Stanford University Press.

Barnes, Barry, and Steven Shapin. 1979. *Natural Order: Historical Studies of Scientific Culture*. Beverly Hills, Calif.: Sage Publications.

Beck, Kurt. 1990. Entwicklungshilfe als Beute. Über die lokale Aneignungsweise von Entwicklungsmaßnahmen im Sudan. *Orient* 4: 583–601.

Bennett, John, and John Bowen, eds. 1988. *Production and Autonomy: Anthropological Studies and Critiques of Development*. Lanham, Maryland: University Press of America/Society for Economic Anthropology.

Berger, Peter L., and Thomas Luckmann. 1966. *The Social Construction of Reality: A Treatise in the Sociology of Knowledge*. Garden City, N.Y.: Anchor Books.

Beverungen, Armin, and Stephen Dunne. 2007. "I'd Prefer Not To": Bartleby and the Excesses of Interpretation. *Culture and Organization* 13 (2): 171–183.

Bierschenk, Thomas, Jean-Pierre Chauveau, and Jean-Pierre Olivier de Sardan, eds. 2000. *Courtiers en développement. Les villages africains en quête de projets.* Paris: Karthala, APAD.

Blau, Peter M. 1956. *Bureaucracy in Modern Society.* New York: Random House.

Bloor, David. 1976/1991. *Knowledge and Social Imagery.* London: University of Chicago Press.

Blumenberg, Hans. 1981/1996. *Wirklichkeiten, in denen wir leben.* Stuttgart: Reclam.

Bonneuil, Christophe. 2001. Development as Experiment: Science and State Building in Late Colonial and Postcolonial Africa, 1930–1970. *Osiris* 15 (second series): 258–281.

Borges, Jorge Luis. 1993. *Other Inquisitions, 1937–1952.* Translated by Ruth L. C. Simms. Austin: University of Texas Press.

Borges, Jorge Luis. 1964. The Analytic Language of John Wilkins. In Jorge Luis Borges, *Other Inquisitions, 1937–1952,* translated by Ruth L. C. Simms, 101–105. Austin: University of Texas Press.

Bourdieu, Pierre. 1992. The Logic of Fields. In Pierre Bourdieu and Löic J. D. Wacquant, *An Invitation to Reflexive Sociology,* 94–114. Chicago: University of Chicago Press.

Braun, Gerald. 1993. Nachhaltigkeit, was ist das? Definitionen, Konzepte, Kritik. In *Hilft die Entwicklungshilfe langfristig? Bestandsaufnahme zur Nachhaltigkeit von Entwicklungsprojekten,* edited by Reinhard Stockmann and Wolf Gaebe, 25–41. Wiesbaden: Westdeutscher Verlag.

Bruner, Jerome. 1986. *Actual Minds, Possible Worlds.* Cambridge, Mass.: Harvard University Press.

Brunsson, Nils, and Johan P. Olsen. 1993. *The Reforming Organization.* London, New York: Routledge.

Buchanan, James M. 1985. *Liberty, Market, and State: Political Economy in the 1980s.* New York: New York University Press.

Callon, Michel. 1986. Some Elements of a Sociology of Translation: Domestication of the Scallops and Fishermen of St. Brieuc Bay. In *Power, Action, and Belief: A New Sociology of Knowledge?,* edited by John Law, 196–233. London: Routledge and Kegan.

Callon, Michel, ed. 1998. *The Laws of the Markets.* Oxford: Blackwell.

de Certeau, Michel. 1980/1984. *The Practice of Everyday Life,* vol. 1. Berkeley: University of California Press.

Clarke, Adele E. 1991. Social Worlds/Arenas Theory as Organizational Theory. In *Social Organization and Social Process: Essays in Honor of Anselm Strauss*, edited by David R. Maines, 119–158. New York: Aldine de Gruyter.

Clifford, James, and George Marcus, eds. 1986. *Writing Culture: The Poetics and Politics of Ethnography*. Berkeley: University of California Press.

Czarniawska, Barbara. 1997. *Narrating the Organization: Dramas of Institutional Identity*. Chicago and London: University of Chicago Press.

Czarniawska, Barbara. 1998. *A Narrative Approach to Organization Studies*. London: Sage.

Czarniawska, Barbara. 2000. *A City Reframed: Managing Warsaw in the 1900s*. Amsterdam: Harwood.

Czarniawska, Barbara, and Bernward Joerges. 1996. Travels of Ideas. In *Translating Organizational Change*, edited by Barbara Czarniawska and Guje Sevón, 13–48. Berlin and New York: Walter de Gruyter.

Czarniawska-Joerges, Barbara. 1993. *The Three-Dimensional Organization: A Constructivist View*. Lund: Studentlitteratur.

Davidson, Donald. 1984/2001. *Inquiries into Truth and Interpretation*. Oxford and New York: Oxford University Press.

DiMaggio, Paul J., and Walter W. Powell. 1983. The Iron Cage Revisited: Institutional Isomorphism and Collective Rationality in Organizational Fields. *American Sociological Review* 48 (April): 147–160.

Douglas, Mary, and Aaron Wildavsky. 1982. *Risk and Culture: An Essay on the Selection of Technical and Environmental Dangers*. Berkeley: University of California Press.

Dubiel, Helmut. 1992. Der Fundamentalismus der Moderne. *Merkur* Sonderheft *GegenModerne. Über Fundamentalismus, Multikulturalismus und Moralisch Korrektheit* (522/523): 747–762.

Duffield, Mark. 2005. Getting Savages to Fight Barbarians: Development, Security, and the Colonial Present. *Journal of Conflict, Security and Development* 5 (2): 1–19.

Easterly, William Russell. 2006. *The White Man's Burden: Why the West's Efforts to Aid the Rest Have Done So Much Ill and So Little Good*. New York: Penguin Press.

Eco, Umberto. 1994. *Six Walks in the Fictional Woods*. Cambridge, Mass.: Harvard University Press.

Edelman, Murray. 1988. *Constructing the Political Spectacle*. Chicago and London: University of Chicago.

Eisenstadt, Shmul N. 1961. Anthropological Studies of Complex Societies. *Current Anthropology* 2 (3): 201–222.

Elshorst, Hansjörg. 1990. Organisation und Entwicklung. Zum System der deutschen Entwicklungspolitik. In *Deutsche und internationale Entwicklungspolitik. Zur Rolle staatlicher, supranationaler und nicht-regierungsabhängiger Organisationen in Entwicklungsprozess der Dritten Welt*, edited by Manfred Glasgow, 19–34. Opladen: Westdeutscher Verlag.

Elwert, Georg. 1987a. Ausdehnung der Käuflichkeit und Einbettung der Wirtschaft, Markt und Moralökonomie. *Kölner Zeitschrift für Soziologie und Sozialpsychologie* (Sonderband 28: *Soziologie wirtschaftlicher Handelns*), 300–321.

Elwert, Georg. 1987b. Die gesellschaftliche Einbettung von Schriftgebrauch. In *Theorie als Passion. Niklas Luhmann zum 60. Geburtstag*, edited by Dirk von Baecker et al., 238–268. Frankfurt am Main: Suhrkamp.

Elwert, Georg. 1996. Kulturbegriffe und Entwicklungspolitik—über "soziokulturelle Bedingungen der Entwicklung." In *Kulturen und Innovationen. Festschrift für Wolfgang Rudolf*, edited by Georg Elwert, Jürgen Jensen, and Ivan R. Kortt, 51–88. Berlin: Duncker and Humblot.

Enzensberger, Hans-Magnus. 1988/1997. Billions of All Countries, Unite! In Hans-Magnus Enzensberger, *Zig Zag: The Politics of Culture and Vice Versa*, translated by Linda Haverty Rugg, 147–185. New York: New Press.

Escobar, Arturo. 1988. Power and Visibility: The Invention and Management of Development in the Third World. *Cultural Anthropology* 3 (4): 428–443.

Escobar, Arturo. 1991. Anthropology and the Development Encounter: The Making and Marketing of Development Anthropology. *American Ethnologist* 18 (4): 658–682.

Escobar, Arturo. 1995. *Encountering Development: The Making and Unmaking of the Third World*. Princeton, N.J.: Princeton University Press.

Ferguson, James. 1990. *The Anti-Politics Machine: "Development," Depoliticization, and Bureaucratic Power in Lesotho*. Cambridge: Cambridge University Press.

Ferguson, James. 2006. *Global Shadows: Africa in the Neoliberal World Order*. Durham, N.C.: Duke University Press.

Foucault, Michel. 1970. *The Order of Things: An Archeology of the Human Sciences*. New York: Random House.

Foucault, Michel. 1978. *Discipline and Punish: The Birth of the Prison*. Translated by Alan Sheridan. New York: Pantheon.

Frankfurter Allgemeine Zeitung. 1997. 19 December.

Fujimura, Joan H. 1992. Crafting Science: Standardized Packages, Boundary Objects, and "Translations." In *Science as Practice and Culture*, edited by Andrew Pickering, 168–211. Chicago and London: University of Chicago Press.

Fuller, Steve. 1994. Toward a Philosophy of Science Accounting: A Critical Rendering of Instrumental Rationality. In *Accounting and Science: Natural Inquiry and Commercial Reason*, edited by Michael Power, 247–280. Cambridge: Cambridge University Press.

Galison, Peter. 1997. *Image and Logic: A Material Culture of Microphysics*. Chicago: University of Chicago Press.

Garud, Raghu, and Peter Karnøe. 2003. Bricolage versus Breakthrough: Distributed and Embedded Agency in Technology Entrepreneurship. *Research Policy* 32: 277–300.

Gellner, Ernest. 1983. *Nations and Nationalism*. Oxford: Blackwell.

Gellner, Ernest. 1992. *Reason and Culture: The Historic Role of Rationality and Rationalism*. Oxford: Blackwell.

Goffman, Erving. 1974/1993. *Frame Analysis: An Essay in the Organization of Experience*. Boston: Northeastern University Press.

Goody, Jack. 1977. *The Domestication of the Savage Mind*. Cambridge, Cambridge University Press.

Goody, Jack. 1986. *The Logic of Writing and the Organization of Society*. Cambridge: Cambridge University Press.

Gould, Jeremy. 2005. Timing, Scale, and Style: Capacity as Governmentality in Tanzania. In *The Aid Effect: Giving and Governing in International Development*, edited by David Mosse and David J. Lewis, 61–84. London: Pluto.

Gourevitch, Philip. 2000. Forsaken: Congo Seems Less a Nation than a Battlefield for Countless African Armies. *New Yorker*, 25 September, 53–67.

Gow, David D., and Elliot R. Morss. 1988. The Notorious Nine: Critical Problems in Project Implementation. *World Development* 16 (12): 1399–1418.

Guattari, Felix. 1994. Praktiken der Zukunft. Modernität und Maschinismus, Technik und Ökosophie. *Lettre Internationale* (spring): 18–21.

Gusterson, Hugh. 1996. *Nuclear Rites: A Weapons Laboratory at the End of the Cold War*. Berkeley, Calif. and London: University of California Press.

Habermas, Jürgen. 1968/1971. *Knowledge and the Human Interests*. Translated by Jeremy J. Shapiro. Boston: Beacon Press.

Habermas, Jürgen. 1993. *Justification and Application: Remarks on Discursive Ethics*. Translated by Ciaran Cronin. Cambridge, Mass.: MIT Press.

Handelman, Don. 1995. Cultural Taxonomy and Bureaucracy in Ancient China: The Book of Lord Shang. *International Journal of Politics, Culture, and Society* 9 (2): 263–293.

Hanke, Stefanie. 1996. Weiß die Weltbank, was sie tut? Über den Umgang mit Unsicherheit in einer Organisation der Entwicklungsfinanzierung. *Soziale Systeme. Zeitschrift für soziologische Theorie* 2 (2): 331–359.

Harper, Richard H. R. 1998. *Inside the IMF: An Ethnography of Documents, Technology, and Organisational Action*. San Diego and London: Academic Press.

von Hayek, Friedrich A. 1948/1996. The Use of Knowledge in Society. In *Individualism and Economic Order*, reissue edition. Chicago: University of Chicago Press, 77–92.

Hegel, Georg Wilhem Friedrich. 1807/2003. *The Phenomenology of Mind*. Translated by James Black Baillie. Mineola, N.Y.: Dover Publications.

Hirschman, Albert O. 1967. *Development Projects Observed*. Washington, D.C.: The Brookings Institute.

Hirschman, Albert O. 1994. Social Conflicts as Pillars of Democratic Market Society. *Political Theory* 22 (2): 203–218.

Hobart, Mark. 1993. Introduction: The Growth of Ignorance? In *An Anthropological Critique of Development: The Growth of Ignorance*, edited by Mark Hobart, 1–29. London: Routledge.

Hoben, Allen. 1982. Anthropologists and Development. *Annual Review of Anthropology* 11: 349–375.

Hobsbawm, Eric. 1995. *The Age of Extremes: A History of the World 1914–1991*. New York: Pantheon.

Holy, Ladislav, and Milan Stuchlik. 1983. *Actions, Norms, and Representations: Foundations of Anthropological Inquiry*. Cambridge: Cambridge University Press.

Horton, Robin. 1967. African Traditional Thought and Western Science. *Africa* 37 (1–2): 50–71.

Horton, Robin. 1988. Tradition and Modernity Revisited. In *Rationality and Relativism*, edited by Martin Hollis and Steven Lukes, 201–260. Oxford: Basil Blackwell.

Horton, Robin. 1993. *Patterns of Thought in Africa and the West: Essays on Magic, Religion, and Science*. Cambridge: Cambridge University Press.

Joerges, Bernward. 1996. *Technik—Körper der Gesellschaft. Arbeiten zur Techniksoziologie*. Frankfurt am Main: Suhrkamp.

Joerges, Bernward, and Terry Shinn. 2001. Research-Technology in Historical Perspective: An Attempt at Reconstruction. In *Instrumentation: Between Science, State and Industry*, edited by Bernward Joerges and Terry Shinn, 241–248. Dordrecht: Kluwer Academic Publishers.

Johnson, Keith. 1985. Conflicting Images and Realities in Project-Planning. *Third World Planning Review* 7 (4): 351–355.

Kalthoff, Herbert. 2007. Ökonomisches Rechnen: Zur Konstitution bankwirtschaftlicher Objekte und Investitionen. In *Zahlenwerk: Kalkulation, Organisation und Gesellschaft*, edited by Andrea Mennicken and Hendrik Vollmer, 143–164. Wiesbaden: VS Verlag für Sozialwissenschaft.

Knorr Cetina, Karin. 1981. *The Manufacture of Knowledge: An Essay on the Constructivist and Contextual Nature of Science*. Oxford: Pergamon Press.

Krappmann, Lothar. 1969/1975. *Soziologische Dimensionen der Identität*. Stuttgart: Klett.

Kühl, Stefan. 2007. Zahlenspiele in der Entwicklungshilfe. Zu einer Soziologie des Deckungsbeitrages. In *Zahlenwerk: Kalkulation, Organisation und Gesellschaft*, edited by Andrea Mennicken and Hendrik Vollmer, 185–206. Wiesbaden: VS Verlag. für Sozialwissenschaft.

Kuhn, Thomas. 1962. *The Structure of Scientific Revolutions*. Chicago: Chicago University Press.

Latour, Bruno. 1986. The Power of Association. In *Power, Action, and Belief: A New Sociology of Knowledge?*, edited by John Law, 264–280. London: Routledge.

Latour, Bruno. 1987. *Science in Action: How to Follow Scientists and Engineers through Society*. Cambridge, Mass.: Harvard University Press.

Latour, Bruno. 1988. The Politics of Explanation: An Alternative. In *Knowledge and Reflexivity: New Frontiers in the Sociology of Knowledge*, edited by Steve Woolgar, 155–176. London: Sage.

Latour, Bruno. 1988/1990. Drawing Things Together. In *Representations in Scientific Practice*, edited by Michael Lynch and Steve Woolgar, 19–68. Cambridge, Mass.: MIT Press.

Latour, Bruno. 1989. Clothing the Naked Truth. In *Dismantling Truth: Reality in the Post-Modern World*, edited by Hilary Lawson and Lisa Appignanesi, 101–126. London: Weidenfeld and Nicolson.

Latour, Bruno. 1990. The Force and Reason of Experiment. In *Experimental Inquiries, Historical, Philosophical, and Social Studies of Experimentation in Science*, edited by Homer Le Grand, 48–79. Dordrecht: Kluwer Academic Publishers.

Latour, Bruno. 1991/1993. *We Have Never Been Modern*. Translated by Catherine Porter. Cambridge, MA: Harvard University Press.

Latour, Bruno. 1995. The "Pédofil" of Boa Vista: A Photo-Philosophical Montage. *Common Knowledge* 4: 144–187.

Latour, Bruno. 1996. On Actor-Network Theory: A Few Clarifications. *Soziale Welt* 47 (4): 369–382.

Latour, Bruno, and Steve Woolgar. 1979/1986. *Laboratory Life: The Construction of Scientific Facts*. Princeton, N.J.: Princeton University Press.

Law, John (ed). 1986. *Power, Action and Belief. A New Sociology of Knwoledge*. London: Routledge.

Law, John. 1994. *Organizing Modernity*. Oxford and Cambridge, Mass.: Blackwell.

Li, Tania Murray. 2007. *The Will to Improve: Governmentality, Development, and the Practice of Politics*. Durham, N.C.: Duke University Press.

Luhmann, Niklas. 1964/1995. *Funktionen und Folgen formaler Organisation*. Berlin: Duncker and Humblot.

Luhmann, Niklas. 1968/1979. Trust: A Mechanism for the Reduction of Social Complexity. In Niklas Luhmann, *Trust and Power*, translated by Howard Davis, John Raffan, and Kathryn Rooney. Chichester: Wiley and Sons.

Luhmann, Niklas. 1969/1973. *Legitimation durch Verfahren*. Frankfurt am Main: Suhrkamp.

Luhmann, Niklas. 1979. *Trust and Power*. Translated by Howard Davis, John Raffan, and Kathryn Rooney. Chichester: Wiley and Sons.

Luhmann, Niklas. 1984/1995. *Social Systems*. Translated by John Bednarz and Dirk Baecker. Stanford, Calif.: Stanford University Press.

Luhmann, Niklas. 1990/1994. *Die Wissenschaft der Gesellschaft*. Frankfurt am Main: Suhrkamp.

Luhmann, Niklas. 1991. Sthenographie und Euryalistik. In *Paradoxien, Dissonanzen, Zusammenbrüche, Situationen offener Epistemologien*, edited by Hans Ulrich Gumbrecht and K. Ludwig Pfeiffer, 58–82. Frankfurt am Main: Suhrkamp.

Luhmann, Niklas. 1993. "Was ist der Fall?" und "Was steckt dahinter?" Die zwei Soziologien und die Gesellschaftstheorie. *Zeitschrift für Soziologie* 22 (4): 245–260.

Luhmann, Niklas. 1997. *Die Gesellschaft der Gesellschaft*. Frankfurt am Main: Suhrkamp.

Lyotard, Jean-François. 1979/1984. *The Postmodern Condition: A Report on Knowledge*. Translated by Geoff Bennington and Brian Massumi. Minneapolis: University of Minnesota Press.

March, James G., and Herbert A. Simon. 1958. *Organizations*. New York: Wiley.

Marcuse, Herbert. 1964. *One-Dimensional Man: Studies in the Ideology of Advanced Industrial Societies*. Boston: Beacon Press.

Martin, Maximilian. 1998. The Deconstruction of Development: A Critical Overview. *Entwicklungsethnologie* 7 (1): 40–59.

Merton, Robert K. 1949/1957. The Self-Fulfilling Prophecy. In *Social Theory and Social Structure*, 475–492. New York: Free Press.

Meyer, John W., and Brian Rowan. 1977/1991. Institutionalized Organizations: Formal Structure as Myth and Ceremony. In *The New Institutionalism in Organizational Analysis*, edited by Walter W. Powell and Paul J. DiMaggio, 41–62. Chicago: University of Chicago Press.

Mosse, David, and David J. Lewis, eds. 2005. *The Aid Effect: Giving and Governing in International Development*. London: Pluto.

Münch, Richard. 1978. Max Webers "Anatomie des okzidentalen Rationalismus": Eine systemtheoretische Lektüre. *Soziale Welt* 29: 217–246.

Münch, Richard. 1984. *Die Struktur der Moderne. Grundmuster und differentielle Gestaltung des institutionellen Aufbaus der modernen Gesellschaften*. Frankfurt am Main: Suhrkamp.

Nabokov, Vladimir. 1957. *Pnin*. Garden City, N.Y.: Doubleday.

Nelson, Benjamin. 1969. *The Idea of Usury: From Tribal Brotherhood to Universal Otherhood*. Chicago: University of Chicago Press.

New York Times. 2004. U.S. Loses Trade Cases and Faces Penalties. 1 September.

Pandolfi, Mariella. 2003. Contract of Mutual (In)difference: Governance and the Humanitarian Apparatus in Contemporary Albania and Kosovo. *Indiana Journal of Global Legal Studies* 10 (1): 369–381.

Pandolfi, Mariella. 2006. La zone grise des guerres humanitaires. *Anthropologica* 48 (1): 43-58.

Petersen, Thomas. 1996. *Individuelle Freiheit und allgemeiner Wille. Buchanans politische Ökonomie und die politische Philosophie*. Tübingen: Mohr.

Pickering, Andrew. 1995. *The Mangle of Practice: Time, Agency, and Science*. Chicago: Chicago University Press.

Pinch, Trevor and Trevor Pinch. 1985. Toward an Analysis of Scientific Observation: The Externality and Evidential Significance of Observation Reports in Physics. *Social Studies of Science* 15: 3–36.

Pinch, Trevor. 1988. Reservations about Reflexivity and New Literary Forms or Why Let the Devil have All the Good Tunes? In *Knowledge and Reflexivity: New Frontiers in the Sociology of Knowledge*, edited by Steve Woolgar, 178–197. London: Sage.

Pinzer, Petra, et al. 1996. Mein Maßstab: "Das Lächeln eines Kindes." Der Chef der größten Entwicklungsbank will mehr als nur Rendite. Ein ZEIT-Gespräch mit James Wolfensohn. *Die Zeit* (22 March): 25–26.

Polanyi, Karl. 1944/2001. *The Great Transformation: The Political and Economic Origins of Our Time*. Boston: Beacon Press.

Polanyi, Karl. 1957/1971. The Economy as Instituted Process. In *Primitive, Archaic, and Modern Economies: Essays of Karl Polanyi*, edited by George Dalton, 139–174. Boston: Beacon Press.

Porter, Theodore M. 1994. Making Things Quantitative. In *Accounting and Science. Natural Inquiry and Commercial Reason*, edited by Michael Power, 36–56. Cambridge: Cambridge University Press.

Porter, Theodore M. 1995. *Trust in Numbers: The Pursuit of Objectivity in Science and Public Life*. Princeton, N.J.: Princeton University Press.

Powell, Walter W. 1990. Neither Market nor Hierarchy: Network Forms of Organizations. *Research in Organizational Behavior* 12: 295–336.

Power, Michael. 1997. *The Audit Society: Rituals of Verification*. Oxford: Oxford University Press.

Quarles van Ufford, Philip. 1988a. The Myth of Rational Development Policy: Evaluation versus Policy-Making in Dutch Protestant Donor Agencies. In *The Hidden Crisis in Development: Development Bureaucracies*, edited by Philip Quarles van Ufford, Dirk Kruyt, and Theodore Downing, 75–98. Tokyo/Amsterdam: United Nations University Press/Free University Press.

Quarles van Ufford, Philip. 1988b. The Hidden Crisis in Development: Development Bureaucracies in between Intentions and Outcomes. In *The Hidden Crisis in Development; Development Bureaucracies*, edited by Philip Quarles van Ufford, Dirk Kruyt, and Theodore Downing, 9–38. Tokyo/Amsterdam: United Nations University Press/Free University Press.

Quarles van Ufford, Philip. 1993. Knowledge and Ignorance in the Practices of Development Policy. In *An Anthropological Critique of Development*, edited by Mark Hobart, 135–160. London: Routledge.

Riles, Annelise. 2001. *The Network Inside Out*. Michigan: University of Michigan Press.

Rorty, Richard. 1979. *Philosophy and the Mirror of Nature*. Princeton, N.J.: Princeton University Press.

Rorty, Richard. 1989. *Contingency, Irony, and Solidarity*. Cambridge: Cambridge University Press.

Rorty, Richard. 1991a. *Objectivity, Relativism, and Truth: Philosophical Papers*, vol. 1. Cambridge and New York: Cambridge University Press.

Rorty, Richard. 1991b. Representation, Social Practice, and Truth. In Richard Rorty, *Objectivity, Relativism, and Truth: Philosophical Papers*, vol. 1, 151–161. Cambridge and New York: Cambridge University Press.

Rottenburg, Richard. 2006. Social Constructivism and the Enigma of Strangeness. In *The Making and Unmaking of Differences: Anthropological, Sociological, and Philosophical Perspectives*, edited by Richard Rottenburg, Burkhard Schnepel, and Shingo Shimada, 27–41. Bielefeld: Transcript.

Rottenburg, Richard, Herbert Kalthoff, and Hans-Jürgen Wagener. 2000. In Search of a New Bed: Economic Representations and Practices. In *Facts and Figures: Economic Practices and Representation (Jahrbuch Ökonomie und Gesellschaft 16)*, edited by Herbert Kalthoff, Richard Rottenburg, and Hans-Jürgen Wagener, 9–34. Marburg: Metropolis.

Sachs, Wolfgang, ed. 1992/1997. *The Development Dictionary: A Guide to Knowledge and Power*. London, N.J.: Zed Books.

Sartre, Jean-Paul. 1943/1956. Bad Faith. In Jean-Paul Sartre, *Being and Nothingness*, 86–118. Translated by Hazel E. Barnes. New York: Philosophical Library.

Sartre, Jean-Paul, Arlette Elkaïm-Sartre, and Jonathan Rée. 1960/1991. *Critique of Dialectical Reason*. London and New York: Verso.

Scherf, Harald. 1992. Fundamentalismus in der Ökonomie. *Merkur* 46 (9/10): 809–819.

Schiffauer, Werner. 1997. Die Angst vor der Differenz. Zu neuen Strömungen in der Kultur- und Sozialanthropologie. In Werner Schiffauer, *Fremde in der Stadt*, 157–171. Frankfurt am Main: Suhrkamp.

Schulz, Manfred, ed. 1997. *Entwicklung. Die Perspektive der Entwicklungssoziologie*. Opladen: Westdeutscher Verlag.

Schütz, Alfred. 1932/1967. *The Phenomenology of the Social World*. Translated by George Walsh and Fredrick Lehnert. Evanston, Ill.: Northwestern University Press.

Schütz, Alfred, and Thomas Luckmann. 1973/1989. *The Structures of the Life-World*. Evanston, Ill.: Northwestern University Press.

Scott, James C. 1990. *Domination and the Art of Resistance: Hidden Transcripts*. New Haven: Yale University Press.

Scott, James C. 1998. *Seeing Like a State: How Certain Schemes to Improve the Human Condition Have Failed*. New Haven: Yale University Press.

Scott, William Richard. 1981/1987. *Organizations: Rational, Natural, and Open Systems*. Engelwood Cliffs, N.J.: Prentice-Hall.

Sennett, Richard. 1998. *The Corrosion of Character: The Personal Consequences of Work in the New Capitalism*. New York: Norton.

Sillitoe, Paul. 1998. What, know natives? Local Knowledge in Development. *Social Anthropology* 6 (2): 203–220.

van der Sluijs, Jeroen, José van Eijndoven, Simon Shackley, and Brian Wynne. 1998. Anchoring Devices in Science for Policy: The Case of Consensus around Climate Sensitivity. *Social Studies of Science* 28 (2): 291–323.

Smith, Barbara Herrnstein. 1997. *Belief and Resistance: Dynamics of Contemporary Intellectual Controversy*. Cambridge, Mass.: Harvard University Press.

Star, Susan Leigh, and James R. Griesemer. 1989. Institutional Ecology, "Translation," and Boundary Objects: Amateurs and Professionals in Berkeley's Museum of Vertebrate Zoology, 1909–1939. *Social Studies of Science* 19: 387–420.

Stengers, Isabelle. 1993/2000. *The Invention of Modern Science*. Minneapolis: University of Minnesota Press.

Strauss, Anselm. 1978a. *Negotiation: Varieties, Contexts, Processes, and Social Order*. San Francisco: Jossey-Bass.

Strauss, Anselm. 1978b. A Social World Perspective. *Studies in Symbolic Interaction* 1: 119–128.

Strauss, Anselm. 1993. *Continual Permutations of Action*. New York: Aldine de Gruyter.

Sülzer, Rolf, and Arthur Zimmerman, eds. 1996. *Organisieren und Organisationen verstehen*. Opladen: Westdeutscher Verlag.

Thompson, Edward P. 1963/1980. *The Making of the English Working Class*. London: Penguin.

Turnbull, David. 2000. *Masons, Tricksters, and Cartographers: Comparative Studies in the Sociology of Scientific and Indigenous Knowledge*. Amsterdam: Harwood.

Waldmann, Peter. 1996. Anomie in Argentinien. In *Argentinien. Politik, Wirtschaft, Kultur und Außenbeziehungen*, edited by Detlef Nolte and Nikolaus Wetz. Frankfurt am Main: Verveurt.

Weber, Max. 1904/1949. "Objectivity" in Social Science and Social Policy. In Max Weber, *The Methodology of the Social Sciences*, edited by Edward A. Shils and Henry A. Finch, 49–85. New York: The Free Press.

Weber, Max. 1904/2002. *The Protestant Ethic and the Spirit of Capitalism*, new translation and introduction by Steven Kalberg. London: Blackwell.

Weber, Max. 1920/1948. Religious Rejections of the World and Their Directions. In Max Weber, *From Max Weber*, translated by H. H. Gerth and C. W. Mills, 267–301. London: Routledge.

Weber, Max. 1923/1927. *General Economic History*. Translated by Frank H. Knight. New York: Greenberg.

Weick, Karl E. 1969/1979. *The Social Psychology of Organizing.* Reading, Mass.: Addison-Wesley.

Weick, Karl E. 1995. *Sensemaking in Organizations.* London: Sage.

Weick, Karl E., and Karlene H. Roberts. 1993. Collective Mind in Organizations: Heedful Interrelating on Flight Decks. *Administrative Science Quarterly* 38: 357–381.

Weiss, Dieter. 1992. Changing Paradigms of Development in an Evolutionary Perspective. *Social Indicators Research* 26: 367–389.

Wittgenstein, Ludwig. 1969. *Über Gewissheit/On Certainty.* Translated by Denis Paul and G. E. M. Anscombe, edited by G. E. M. Anscombe and G. H. von Wright. New York: Harper.

World Bank. 1993. *Getting Results: The World Bank's Agenda for Improving Development Effectiveness.* Washington, D.C.: The World Bank.

Zammito, John H. 2004. *A Nice Derangement of Epistemes. Post-positivism in the Study of Science from Quine to Latour.* Chicago and London: Chicago University Press.

Index

Access (to the field), xxxiv, 59–61
Accountability, 73, 182
  commercial, 82
  trap, 152, 167
Actor network, 125
Agency, distributed, xii, xxvi, 117
Agnosticism, xxxiii, xxxiv
Arena, xii
Artifact, xxi, xxxi, xxxii, 80, 104
Avoidance strategy, 114–116, 141

Black box, 26, 60, 61, 65
Blind spot, xxviii, 152, 200
Blueprint approach, 73, 96, 142, 149,
  196. *See also* Standardized package
Boundary object, xxvi, 104, 212n5
Budget analyses (impact of), 110
Bureaucracy, 27, 67, 76, 140–141, 166

Calculation, xxvii, 36, 101, 108, 148,
  215n3
Center of calculation (taxonomy),
  87–88, 181–187, 190
Classification system, 137
Code switching, 198
Commitment (to the rules), 62–65, 128
Common good, 98, 109, 166, 175
Communication, 78, 125, 151, 207n10
  defensive, 85, 210n28
Consensus of experts (as authorization
  of truth claims), 79

Constructivism, 173, 209n21, 214n2
Consultant (or contractor), 102–104
Contract, 32–33. *See also* Noncontractual
Cooperation, 104, 149–151, 174, 191–
  194, 198–200. *See also* Interface
Coordination (of development projects), 13, 43, 69, 114–115, 117, 144,
  213
Correspondence theory, xxix–xxii,
  76, 92, 108, 111, 139, 151, 183–184,
  189–190, 199–200
Cost-benefit analysis, 99
Cultural code, xxix, 77, 173, 198–200
Cultural difference, 130, 136, 159, 180,
  196–197, 214n1
Culture (as something others have),
  xiii, xxvi, xxxi, xxxiv, 60–61, 68, 73,
  75, 80, 88, 104, 108, 111, 130, 142,
  144, 198–199
Culture of writing, 108, 142

Data (valid, reliable, consistent),
  127–128
Data, elementary, or aggregated. *See*
  Figures
Databank, 126–130
Deconstruction, xxiv, 173, 203–204n15,
  208–209n21
Defense mechanism (or defense strat-
  egy). *See* Avoidance strategy

Inside Technology

edited by Wiebe E. Bijker, W. Bernard Carlson, and Trevor Pinch

H. M. Collins, *Artificial Experts: Social Knowledge and Intelligent Machines*

Paul N. Edwards, *The Closed World: Computers and the Politics of Discourse in Cold War America*

Herbert Gottweis, *Governing Molecules: The Discursive Politics of Genetic Engineering in Europe and the United States*

Joshua M. Greenberg, *From Betamax to Blockbuster: Video Stores and the Invention of Movies on Video*

Kristen Haring, *Ham Radio's Technical Culture*

Gabrielle Hecht, *The Radiance of France: Nuclear Power and National Identity after World War II*

Kathryn Henderson, *On Line and On Paper: Visual Representations, Visual Culture, and Computer Graphics in Design Engineering*

Christopher R. Henke, *Cultivating Science, Harvesting Power: Science and Industrial Agriculture in California*

Christine Hine, *Systematics as Cyberscience: Computers, Change, and Continuity in Science*

Anique Hommels, *Unbuilding Cities: Obduracy in Urban Sociotechnical Change*

Deborah G. Johnson and Jameson W. Wetmore, editors, *Technology and Society: Building our Sociotechnical Future*

David Kaiser, editor, *Pedagogy and the Practice of Science: Historical and Contemporary Perspectives*

Peter Keating and Alberto Cambrosio, *Biomedical Platforms: Reproducing the Normal and the Pathological in Late-Twentieth-Century Medicine*

Eda Kranakis, *Constructing a Bridge: An Exploration of Engineering Culture, Design, and Research in Nineteenth-Century France and America*

Christophe Lécuyer, *Making Silicon Valley: Innovation and the Growth of High Tech, 1930-1970*

Pamela E. Mack, *Viewing the Earth: The Social Construction of the Landsat Satellite System*

Donald MacKenzie, *Inventing Accuracy: A Historical Sociology of Nuclear Missile Guidance*

Donald MacKenzie, *Knowing Machines: Essays on Technical Change*

Donald MacKenzie, *Mechanizing Proof: Computing, Risk, and Trust*

Donald MacKenzie, *An Engine, Not a Camera: How Financial Models Shape Markets*

Maggie Mort, *Building the Trident Network: A Study of the Enrollment of People, Knowledge, and Machines*

Peter D. Norton, *Fighting Traffic: The Dawn of the Motor Age in the American City*

Helga Nowotny, *Insatiable Curiosity: Innovation in a Fragile Future*

Ruth Oldenziel and Karin Zachmann, editors, *Cold War Kitchen: Americanization, Technology, and European Users*

Nelly Oudshoorn and Trevor Pinch, editors, *How Users Matter: The Co-Construction of Users and Technology*

Shobita Parthasarathy, *Building Genetic Medicine: Breast Cancer, Technology, and the Comparative Politics of Health Care*

Trevor Pinch and Richard Swedberg, editors, *Living in a Material World: Economic Sociology Meets Science and Technology Studies*

Paul Rosen, *Framing Production: Technology, Culture, and Change in the British Bicycle Industry*

Richard Rottenburg, *Far-Fetched Facts: A Parable of Development Aid*

Susanne K. Schmidt and Raymund Werle, *Coordinating Technology: Studies in the International Standardization of Telecommunications*

Wesley Shrum, Joel Genuth, and Ivan Chompalov, *Structures of Scientific Collaboration*

Charis Thompson, *Making Parents: The Ontological Choreography of Reproductive Technology*

Dominique Vinck, editor, *Everyday Engineering: An Ethnography of Design and Innovation*